Europe-Asia-Pacific Studies in Economy and Technology

Theodor Leuenberger (Ed.)

From Technology Transfer to Technology Management in China

With 6 Figures

Springer-Verlag Berlin Heidelberg New York
London Paris Tokyo Hong Kong

Prof. Dr. Theodor Leuenberger
Institut für Technologiemanagement
Unterstraße 22
CH-9000 St. Gallen, Switzerland

This publication was supported by a grant from the Volkswagen-Stiftung as well as by a gra
Research Fund of the University of St. Gallen.

ISBN-13:978-3-642-75635-1   e-ISBN-13:978-3-642-75633-7
DOI: 10.1007/978-3-642-75633-7

2142/7130-543210

Preface

Access to science and technology worldwide is achieved by active participation in open international scientific research, as well as through technological capability that is decisive in catching up with world developments in science and technology. In other words, it is the "national system of innovation" which determines a country's assimilation capacity. The universities, research institutions, the technological infrastructure, industrial training schemes, information networks and technical institutions in general provide the foundation for a solid, steady development.

Therefore policies directed toward strengthening the national system of innovation are essential for a catching-up strategy. But even more important is the presence of skilled and experienced people with the necessary connections to the scientific and technological infrastructure of the world at large.

this applies to China in particular. Whether or not the technological potential will be developed, depends on the technological and industrial strategies promoted by the Chinese leadership.

In addition, the costs and benefits of technological development are affected by the prevalent evolutionary stage of a country's political framework and fiscal regime. There must be a strong coordination between overall economic policies and technology policy. A sucessful management of technology is only possible through a "technological package" including management, financial and marketing skills.

* * *

I would like to thank numerous friends and colleagues, whose kind assistance has made this publication possible. I am particularly grateful to my colleagues in the various research policy institutions in Beijing.

Part Three of this publication consists of selected contributions from the International Conference organized by the Development Research Centre of the State Council from May 30 untill June 2, 1988 in Beijing on the initiative of Martin Lees, Paris. These contributions are reprinted herein with the kind permission of Mr. Ma Hong and Wu Mingyu, Directorate State Council, Development Research Centre, Beijing.

Thanks are also due to the staff of UNDP in Beijing, since it was in the context of UNDP activities that we were offered many opportunities to study key development problems in the field of science and technology. I would further like to thank the members of my staff at the Institute for Technology Management, who have not only created a supportive environement but provided me with the necessary human skills and resources.

Theodor Leuenberger
February 1990

Contents

Preface: Theodor Leuenberger   V

Part One: Technology Management

Deng´s Reforms 1976-1988   3
Manfred Kulessa, Berlin

Domestic and Foreign Technology
Factors Influencing Assimilation and
Diffusion Capabilities   14
Richard Conroy, Paris

From Technology Transfer to Technology
Management   72
Pierre Ventadour, Paris

Changing Chinese Thinking about Technology
Transfer*   102
Ryusuke Ikegami, Tokyo

Part Two: Case Studies

The Development of the Chinese Steel Industry   125
Jaques A. Astier, Paris

Technology Transfer in China: The Case of
Oxygen-Generating Equipment in Steel Industry,
1978-1988   154
Eduard B. Vermeer, Leiden

Part Three: China and the World in the Nineties

A Summary of Global Technology Trends of
Possible Strategic Interest to the
People´s Republic of China   211
Lewis M. Branscomb, Cambridge MA

China and the World in the Nineties,
Trends in New Technologies and Their Implications
for China toward the 1990s   227
Keichi Oshima, Tokyo

China´s Strategy for Agricultural Development
in the 1990s   233
He Kang, Beijing

Scientific and Technological Progress
and the Revitalization of China´s Economy   242
Hu Ping, Beijing

China and the World in the Nineties,
Deepening Reform for Technological Progress
in China   251
Lin Zixin, Beijing

The Rapid Expansion of Economic Information
in the 1990s and the Challenge to China´s
Economic Reform   261
Zhou Xiaochuan & Yang Jianhua, Beijing

Contributors   283

Part One

Technology Management

DENG'S REFORMS 1976-1988

Manfred Kulessa
Entwicklungspolitisches Forum
Deutsche Stiftung für internationale Entwicklung (DSE)
Reiherwerder, D-1000 Berlin 27

As all policy in China, the reforms, too, are undergoing a time of crisis. The Chinese term for "crisis" (wei-ji) is composed of the characters for "danger" and "opportunity". If we wanted to describe the state of crisis, we would have to identify elements of danger and elements of chance and hope. Our Chinese friends, who like to think in terms of history, perceiving the world in images of dynamic evolution, are usually very conscious of the connection between crisis and change. On the other hand, we can also try to recognize elements of continuity and of change, as Western historians like to do. Perhaps it is still too early to allow for a thorough analysis of the events of 1989. But one can certainly review the dangers and chances of the reforms designed and enacted in the years before.

In those years, we sometimes started with a humorous question: "Each Chinese has one, but the whole of China knows only twelve - what is it?" Of course, the reference is to the Chinese zodiac, which is composed of animal signs returning within a cycle of twelve years. One such cycle stood under the sovereign leadership of Deng Xiaoping: 1976-1988. Deng himself told us that, in his judgement, twenty of the People's Republic's first forty years went wrong. He saw positive developments only in the first and fourth decades. Deng was born in the year of the dragon; and 1976 turned out to be a typical dragon year, a wild year, a year of crises. Zhou Enlai and Mao Zedong died. The earth trembled. In Tanshan alone, the quake killed about half a million people. In times of grave danger, the chance for a major change occurred. The Gang of Four was dethroned, the Cultural Revolution finished, and the reform policy initiated. The Middle Kingdom had a new ruler.

Europe-Asia-Pacific Studies in Economy and Technology
Leuenberger (Ed.) From Technology Transfer
to Technology Management in China
© Springer-Verlag Berlin Heidelberg 1990

The celebration of the 35th anniversary of the People's Republic on October 1, 1984, became the symbolic climax of this rule, when one man alone went out of the Gate of Heavenly Peace to meet the troops, respectfully watched by all the dignitaries and envoys, and by millions of onlookers. In 1988, again a dragon year with all kinds of natural catastrophes, the old helmsman held out in order to complete the task of reconciliation with the Soviet Union. When he finally succeeded in May 1989, the earth had trembled again, and the student protest in the Square of Heavenly Peace already indicated the next crisis: danger and opportunity again. The following comments address the reforms of Deng's twelve years in the cycle of 1976 to 1988. They deal with the policy as expressed in the resolutions of the Third Plenary of the 11th Central Committee of December 1978 and the following decade.

Let us remember the difficult starting conditions of 1976. In addition to the massive development problems of a very large and poor country, the leaders had to face some serious obstacles on the way to a new beginning. I would like to mention just six of them: the heritage of the Cultural Revolution, the population problem, the low level of industrial productivity, bottlenecks in transport and energy, the East-West discrepancy and, finally, the question of modern ethics.

As we know, most Chinese find it difficult to overcome the experiences of the Cultural Revolution. It has resulted in traumatic memories and, at the same time, in a huge national gap in education. German observers are inclined to compare the Chinese trauma with our problems of dealing with the memories of the terrible twelve years of the Hitler period. Indeed, we may be able to explain certain phenomena in this way: how sympathizers could become activists, for instance, and activists victims. Still, the situation is quite different. In Germany it was much easier to claim that everything had changed with the end of the lost war. From fascism, we had turned to democracy. In China, one has to deal with the same Party, often enough with the same functionaries, and certainly with the same colleagues, neighbors and relatives whose behavior during these critical years is still fresh in one's memory. With the exception of a few who were involved in brutal crimes, all are still together and have to get along with each other. This is not always easy to stomach. On the national level, the problem was partly solved by focussing on the so-called Gang of Four, and using them as a scapegoat. Nevertheless, a serious psychological problem has remained. In many biographies, the deviation has not been smoothed out. Again, similarly to the years of the Group of 47 in Germany, the literature of the post-Cultural-Revolution period has largely been absorbed by the task of coming to terms with the ugly past.

It was not easy for the reformers to have to tackle the population problem simultaneously with the reforms. Against Chinese tradition and the general trend of liberalization, they had to insist on discipline and reglementation in a country of intimately personal character. But there was no alternative: population policy, population and family planning had to be enforced. No progress could be conceived of without a reduction in population growth. In fact, China's population has doubled in the past forty years: there were 540 million Chinese in 1949; today, there are more than 1.1 billion. To feed a billion people and to provide adequate work and income for them, presents an immense challenge in itself. Therefore, planning figures were established. At the end of the century, the population was to stand at 1.2 billion (today, the figure mentioned is 1.3 billion) and to decrease later, finally to stabilize around the middle of next century. With such targets in mind, a policy was designed that was to be realized with as much education, persuasion and voluntary cooperation as possible, but also with as much pressure as necessary.

A different but also tough practical problem of serious economic dimensions presented itself in the low level of industrial productivity. For a layman in the field of economic statistics, it is not easy to demonstrate this on a national scale. It may not even be possible to compare productivity at all, as some scientists have advised, while others are willing to express the view that China's industrial production could, in quantitative terms, be compared with that of Great Britain. Obviously, the input is much more extensive in China. Something has to be changed if China wants to achieve progress. The problem has technical, personnel and institutional components: all three need tackling.

In the field of energy and transport, central bottlenecks had to be faced. The existing and constantly growing demand could not be met, as the vast investment required could not be made in time. The first stretches of real motor highway are only now being built. Shanghai, the main economic center, has been promised a subway. The railroads cannot cope with long-distance transport. Users face constant competition for passenger cars and seats, as well as for time allotments. Equally, energy is in short supply and has to be rationed in many places, a fact that contributes to low productivity in industry and to hardships among people. Huge coal deposits ensure the supply for centuries. But there are also the problems of ecology and of transport. A mega-project on the Yangtse promises relief: the dam at the Three Gorges would be the world's largest producer of hydropower, and its location would suit Chinese industry well. The "project of the century", whose cost is estimated at between twenty and thirty US dollars, is subject to a heated debate in academic and political circles. While environmental aspects and the problem of relocating the local population are also being discussed, the main question appears to be whether it can be justified to put so many eggs into one basket, to concentrate so many resources on one single project. As the economic advantages in terms of energy and transport appear rather convincing, it is likely that the final decision will be taken in favor of this project. The dam will be built, though probably only next century.

Then, the East-West discrepancy. Just as the world knows a North-South divide between rich and poor, between the more or less industrialized and the developed countries, one may speak of an East-West contrast in China. In the coastal areas, the former regions of colonial domination and influence, progress is obvious. There, people actively participate in the changes brought about by reforms, seek and find access to external markets, and are successful in their efforts. In comparison, it looks much bleaker in the hinterland, often inhabited by national minorities, in the remote regions and mountain ranges, and in the old revolutionary bases and areas visited by the Long March (a fact with some political significance). In these areas, people are often unable to make ends meet and have to rely on relief assistance. Of course, these areas are usually not heavily populated. If one divides China diagonally by a line running from northeast to southwest,

only about six per cent of the population live in the western half,
which also includes uninhabited desert and mountain regions. But we
notice an increasingly political debate about the development of the
Western provinces, as Chinese opinion seems to be divided when slogans
such as "Let us build four, five Hong Kongs in China!" are mentioned.
To some, this describes an important economic target, to others a
vision of horror.

When Henry Kissinger called on Deng Xiaoping, many foreigners wanted
to know which problems of China they had discussed. Kissinger said
laconically: "Population, pollution, corruption." I still do not know
whether this is an original quotation from Deng or from Kissinger. In
any case, all three problem areas are related to ethical questions.
There is a need to build a new ethical foundation for the reform era.
This need is felt by the man in the street as well as by the intellectual elite, which is following a trend in research in ethics. When
addressing Parliament in the spring of 1988, Zhao Fusan, the Vice-Pre-
sident of the Academy of Social Sciences, did not only point out this
particular need, however; he also recommended looking at religion for
useful elements of such a new ethical foundation. This was nothing
short of a sensation. Certainly, religious freedom is tolerated by
the reform constitution of 1982. But it was a totally uncommon concept that religion had something to offer that could usefully be
learned and adopted in modern society. The religions practiced in
China are Buddhism, Islam, Christianity and Taoism. Only Taoism is of
Chinese origin, but does not have too much actual importance at present; the other three entered China from the outside, and their current influence appears to be limited to the rather small groups of
their active followers or to some national minorities (Tibet, Xinjiang). On the whole, Confucianism appears much more relevant. Confucianism, however, is usually described, not as a religion, but as an
ethical-philosophical system. It was formulated twenty-five centuries
ago in the historical situation of the feudal period, emphasizing the
ideal of a harmonious balance of spirit and behavior, human virtues,
and the duty to respect and honor elders and superiors. All this was
subject to attack in the Cultural Revolution, as Zhou Enlai himself
became the target of the so-called anti-Confucius debate. In the
meantime, a Confucius Federation has been founded again, trying to de-

fine what Confucius has to offer in modern times. As we know, the whole region of East Asia has been culturally shaped by Confucianism. Not only in Taiwan or Singapore but even in the PRC do we notice an effort to rediscover and represent lasting values. Obviously, much has to be reformed. On quoting Confucius, at least his views regarding the status of women and the meaning of law must be criticized as definitely outdated. Nevertheless, we still find ourselves faced with the fascinating question whether Confucian attitudes and socialist creed can be combined into a new system of ethics.

The basic resolutions of the reform policy were adopted in December 1978. Two topics moved into the foreground: one the one hand, modernization and an opening to the outside world were postulated; on the other hand, targets for economic growth were identified. The economy was expected to quadruple within the last two decades of this century. (At the end of the first decade, it looks as if this target can be realistically achieved, taking into account the low starting level and the growth rates recorded so far.) By the year 2050, China was expected to reach a median level, a position of relative prosperity. When mentioning such terms, the Chinese do not think of the Japanese example, but rather of other Asian neighbors who are somewhat better off than the mainland Chinese themselves.

Modernization was announced in four areas: agriculture, science/technology, industry, and defense. Here, access to modern development, i.e. mainly Western technology, was to be gained. Such modernization, as one knows in China, too, has to do with preconditions and consequences. Such a type of modernization as has been desired is not conceivable without modern information and communication systems, a functioning system of law and order, and a degree of democracy. The policy of opening mainly means international trade, foreign investment and technology transfer. Occasionally, China's leadership showed some concern about the by-products of such opening policies. One used to say: "We open the window, but we do not want to get all your flies; therefore we keep some kind of mosquito net in place." Twice, there were campaigns: in 1983 against "spiritual pollution", and in 1986-87 against "bourgeois liberalization". Even earlier, dissidents such as Wei Jingsheng and Xu Wenli had been sentenced to long prison terms. They had requested democracy as a fifth type of modernization.

Beginning in 1978, rural reforms were tackled immediately and with great resolve. The communes were dissolved, and the farmer families received land under the so-called household responsibility system, providing them with a kind of long-term lease which allowed them to work the land on their own, while obliging them to sell a part of their harvest to the state at fixed prices. This reform turned out to be very successful. It freed dynamic forces and brought about growth and diversification in agricultural production. The importance of this development can hardly be overestimated. After all, four fifths of the population live in rural areas, and although China has only seven per cent of the world's arable land, it has to feed twenty-two per cent of the world population. There is also an interesting strain of historical continuity. Traditionally, a change in dynasty used to start with a movement from the countryside. Thus, Mao built his re- volution on the support from poor peasants, and Deng again introduced himself through an act of liberating farmers. This was tremendously popular. I have been told that in some villages, farmers only took one night of bargaining and discussion to divide the land among them- selves. Deng was correct in assuming that Chinese farmers resented the high degree of collectivism forced on them. But one can also de- tect elements of continuity in the succession of Mao, who wanted to strengthen the farmers' buying power, leave the utilization of local resources to the local people, and who saw the role of the state in monitoring and directing rather than in assuming command in questions of practical implementation. The success of rural reforms could be seen in terms of increased agricultural production and in a rapidly increasing industrialization of the countryside. Grain production reached its peak in 1984 with a total of 407 million tons. Since then, it has been stagnating and is even in a slight decline. By now, we can recognize limitations, for instance as a consequence of ecolo- gical burdens, energy waste, and a lack of cooperation between small production units - a type of cooperation that has to be learned again and again. A major quantitative increase in agricultural production could only be achieved through massive additional investments. This

issue has been the subject of numerous discussions among many responsible officials and leaders in China. In any case, it is inconcei-
vable to feed the masses of China through any type of large-scale imports. In fact, China still has the lead in the production of three
of the four main plants that feed mankind: 37 per cent of the rice is
harvested in China, as is 31 per cent of the world's potato and sweet
potato crops and 17 per cent of the wheat (in which China equals the
production of the US and Canada). Only in corn does the United States
have a leading position: 46 per cent, as against 14 per cent in China.

Since the middle of the reform decade, urban industrial reforms have
been the center of attention. October 20, 1984, may be regarded as
the decisive date in this respect: on that day, the Third Plenary of
the 12th Central Committee passed a resolution on the economic system,
which mainly referred to industry. The major industries had been led
and managed by the state ever since liberation. At government level,
a separate ministry had been established for each branch of industry.
Central planning had been the main instrument. Profits had gone back
to government accounts, losses had simply been recorded. Now, the
system was to be changed. In a socialist market economy, the state
was to regulate the market, which in turn was to guide production.
Thus, management became important, and productivity a real target.
Quality control was introduced. The enterprise received a new central
purpose and aim: to make profits and to pay taxes. The state remains
the owner of the major industries but is now trying to allow management to develop its own skills. This is the core of a policy which
was then formulated in the Enterprise Law of 1988, in the much-debated
Bankruptcy Law, in the labor laws, and in the draft for a social security system.

Contrary to the rural reforms, popular reaction was not only characterized by enthusiasm but also by many fears. There was the fear of
whole factory crews that their unit might be in danger of bankruptcy,
the individual staff member's fear of losing his job for lack of qualifications or for reasons of production technology. Even more, pen-
sioners had been able to expect to place one of their children in
their factory or workshop; the factories as responsible units (danwei)
had been the source of everything necessary, providing housing, health

care and an old-age pension. At the same time as the iron rice bowl
was destroyed, this whole system was put into question. A certain
feeling of insecurity followed. New systems of providing security had
to be created, for instance through the introduction of a general
social security system. There was also a fear of inflation. We must
remember that the large majority of contemporary Chinese had been un-
familiar with the phenomenon of accelerating inflation. Now, prices
were supposed to play a regulatory role. The first price deregula-
tions led to an inflation rate of twenty per cent and more. Hard-
boiled economists may not be frightened by such figures, but for the
urban population of China, they came as a shock, all the more so since
these developments in fact meant a reduction in living standards for
about one fifth of the citizens. All this led to a good deal of unrest among the city people in recent years, as well as to the habit of
investing savings in consumer goods as soon as possible. There was
also some confrontation and unrest in enterprises, even strikes -
which should be ruled out if one adheres to the 1982 Constitution of
the PRC. I also believe that the revolt of the students, themselves
children of the cities, and its support by the population, has something to do with these basic fears and the general unrest, although
there certainly were other motives and incidents that contributed to
the start of the demonstrations and to their prolonged course of
action.

If a government is seriously determined to reduce its role from economic management to economic policy direction, then the necessary steer-
ing instruments for such policies must not only exist, but function
well. They include a strong if not independent reserve bank, a banking system where credit is subject to economic criteria, a reformed
tax system, a system of economic information, the development of a
housing market, an administrative reform with a good deal of authority
delegated to provinces and towns, adequate training or retraining of
managers and officials, a new type of economic planning, the implementation of factory democracy, and a new role for the trade unions.
If the Party is to refrain from directly influencing production and
administration, far-reaching political reforms are required which have

to include a reform of the legal system geared toward the rule of law,
and a process of democratization with basic changes in the election
system.

What could only be outlined here, may indeed by identified as a co-
lossal task of reform, as it was called by Zhao Ziyang in his grand
report at the 13th National Congress on October 25, 1987. This report
is the key document of the reforms of the Deng period. It was published under the slogan, "Advance along the Road of Socialism with
Chinese Characteristics." The report describes the reforms in their
political and economic dimensions and underlines a basically pragmatic
and open-minded attitude, repeating the slogan, "Liberate the spirit,
seek truth from facts, look jointly into the future!" But it also
mentions the borderline which was drawn by Deng Xiaoping when he announced the so-called four basic principles ten years ago, including
the absolute leadership of the Communist Party. We were intrigued by
the reasoning. "Why," asked Zhao, "must we adhere to the four cardi-
nal principles? Because in contemporary China, this is the only way
we can fully guarantee the growth of the productive forces." Actually, the Chinese reforms had already reached a difficult stage of
transition in 1987-88. Elements of the new and old systems existed
side by side. The reform protagonists looked all over the world for
useful examples and options. They wanted to learn how to move resolutely, but in a step-by-step fashion, without causing too much unrest
and harm in the transitory stages. In these circles, one thought then
that nobody could afford a radical break.

Almost ten years ago, the Norwegian peace researcher, Johan Galtung,
put forward his thesis that the history of the People's Republic was
characterized by an oscillation between phases of distributory justice
and growth orientation, and that the period of Deng's reforms would be
followed by a phase of more emphasis on distribution. Whether this
will happen is hard to say. In the long run, we think it more likely
that the Chinese will move forward, with growing success, to combine
dynamic elements of the market economy with elements of socialist responsibility into models of uniquely Chinese structure. We also think
it most urgent to meet the challenge of "one country - two systems",
not only in the sense of integrating Hong Kong. Zhao said that he

would need another two to three years to master the crisis of change. Now, after his fall and in the face of the situation as it presents itself after June 1989, directions and time frames will have to be re-assessed.

We recall Napoleon's famous phrase that the world will tremble when the giant China awakes. In our time, there is no doubt that the giant is going to awake. Sometimes, we may have reason to tremble. But the main development lies in the fact that China is bound to become an increasingly important factor in world affairs, and an equally important partner for Europe. Therefore, we have a vital interest in seeing this great country master its crises and utilize its opportunities.

DOMESTIC AND FOREIGN TECHNOLOGY -- FACTORS INFLUENCING ASSIMILATION AND DIFFUSION CAPABILITIES*

Richard Conroy

OECD Development Centre

94 rue Chardon-Lagache, 75016 Paris

It is nowadays a truism to say that technology is the key to economic development. While the exact contours are as yet unclear, the impact of the so-called new technologies in particular is having and will increasingly have a profound effect on all economies, whether or not individual countries participate in any way in their generation. As access to and control over technology is becoming the key to industrial competition both domestically and internationally, policies aiming at gaining access to technology, or exploiting its full potential if it is already in use, are high on the list of priorities of both industrially advanced and developing countries. This is at a time when new actors have been integrated into the global economy, while traditional actors have seen their competitiveness eroded. The success of the former has in no small measure been due to their ability, in a favourable international economic climate, to acquire technology from the latter, assimilate it rapidly and efficiently, and utilise it to promote technological and industrial capabilities. A response of the traditionally dominant actors has been to institute neo-mercantilist strategies which, on the one hand, attempt to promote and protect their technological superiority, while on the other hand they try to open up foreign markets, particularly those with future growth potential

---

* The material for this chapter was taken from a more detailed study, "Technological Change and Industrial Development in China", currently being prepared by the author. The views expressed are the author's alone and do not necessarily reflect those of the OECD.

Europe-Asia-Pacific Studies in Economy and Technology
Leuenberger (Ed.) From Technology Transfer
to Technology Management in China
© Springer-Verlag Berlin Heidelberg 1990

in Latin America and Asia. The growing politicisation of international trade and investment and the more important role played by governments in securing technological advantages are being played out, not only at national levels but increasingly at regional levels, as attempts to extend market boundaries coincide with steeply escalating costs of R&D and innovation, especially for the so-called generic technologies.

It is in this complex, uncertain and rapidly changing external environment that China has, in the past ten years, started tentatively to integrate itself more into the global economy through expanded trade and investment, and to gain greater access to technology to improve its currently poor international competitiveness. That this access is crucial to China's overall modernisation plans is without doubt, given the past and current problems of generating the required technology domestically. The fact that China possesses a comprehensive but technologically backward industrial base means that because of foreign exchange constraints, foreign technology can only provide a relatively small part of its total needs. However, Chinese commentators argue that a selective acquisition strategy could, under given conditions, have a significant impact on industrial development through both a "knock-on" and a "leapfrog" effect. It is further argued that technology imports could have a significant effect on improving domestic technological capabilities, leading to a "virtuous cycle" development strategy whereby imported technology is mastered and applied, modified and improved. This roughly corresponds to the model proposed by Dahlman et al. (1) to explain the success of certain NICs in managing their technological development by combining domestic and foreign technology components in order progressively to build up production, investment and innovation capabilities (i.e., capabilities to operate and maintain efficient production with imported equipment, to duplicate and modify the equipment, and independently to design, manufacture and improve upon the original imported equipment).

In attempts to upgrade its technological base and narrow the gap not only with industrially advanced countries, but also with some of its

Asian neighbours, the Chinese authorities face a number of options
and constraints in importing technology. These include the question
of access to advanced, particularly to the so-called dual-use,
technologies, the volume and sources of investment, the form of
transfer, and the appropriate sectoral priorities. Such variables
are a function of both internal and external conditions. Of greater
long-term importance, however, is the degree to which assimilative
capacity can be enhanced in order to upgrade domestic technological
capabilities and derive maximum economic benefit from the investment
in imported technology. This depends critically not only on
goverment policies but also on the specific response of production
enterprises to such policies. Failure to develop a strong
assimilative capacity will result in increasing reliance on purely
domestic resources as a source of new technology, dependence on
continual imports of technology, combined with less ability to
select and master them, and a widening of the technological gap.
This scenario has clear ramifications for China's international
competitiveness, especially vis-à-vis its competitors, and by
extension for its export earnings and its consequent ability to
finance future imports of technology, equipment and materials needed
in its ambitious modernisation strategy. Difficulties of access to
new technology and poor assimilation capabilities could also
strengthen xenophobic tendencies, never far from the surface, and
push China back into a more autarkic posture, affecting trade and
investment flows, especially with industrially advanced countries,
to the detriment of all parties concerned.

The current task faced by the Chinese authorities therefore is to
ensure the maximum utilisation of its existing S&T resources,
acquire foreign technology selectively and meld the two in such a
way as results in both efficient production and a gradual increase
in domestic technological capabilities. Such management is complex
and interactive, as the ability to evaluate and select the foreign
technology most appropriate to Chinese conditions and industrial
development strategy is partly a function of its existing
technological capabilities, as is its effective assimilation.
However, under certain circumstances, imported technology may result
in the marginalisation of the domestic S&T system and contribute

little to developing new technological capabilities. The efficient management and utilisation of domestic S&T resources is therefore a key factor in ensuring that China's technology acquisition policy objectives are achieved.

Based on this background, this chapter examines the following issues. A brief survey of technological levels in China is followed by a discussion of the constraints faced in developing technology domestically and in transferring it to the industrial sector. A broad overview of the role of technology acquired from abroad is then presented and the major problems that have been met in assimilating this technology are analysed.

Technological levels and the domestic generation and application of technology

The change in economic development strategy in the late 1970s from an extensive to a more intensive growth pattern forced decision-makers to reassess the role technology had played in industrial growth in the previous thirty years. It was conceded that past output growth, although high at 8.7 per cent per year between 1957-83 had been largely due to an equal or faster rate of growth of inputs of labour and especially capital. As Tidrick has observed, the rapid growth of the latter in comparison with both labour and output has left China with slow employment growth and a rising capital-output ratio (2). Without more efficient use of inputs, the maintenance of industrial output growth can only be achieved by ever-increasing investment rates. The previous high rates of investment should have accelerated factor productivity growth by embodying improved technology in new plant and equipment, but this has not happened. The evidence for widespread technological stagnation is clear and unequivocal, and the Chinese authorities have identified an extensive upgrading of industrial technological levels as the key to its industrial modernisation strategy, with a number of ambitious targets to be reached by 1990 and the year 2000. Although this strategy is essential for improving poor productivity levels, it is of course not the only way of improving factor productivity. Resources can be relocated to higher productivity sectors, labour and management quality can improve, greater economies of scale could be achieved, better efficiency in using existing technology could be realised, as well as changes in the techniques of production. However, activities such as reducing costs, introducing new products and processes, improving quality and productivity all directly or indirectly involve issues of access to technology, its assimilation and diffusion. As Tidrick has concluded, the key to instituting an intensive growth pattern in China lies in both access to modern technology combined with a set of systemic reforms which will generate incentives for enterprises to make sound investment decisions and pay constant attention to costs, and product quality.

A country's technological level is very difficult to measure but there is a clear consensus in China that for many years, little attention has been paid to introducing new products and processes or even improving old products (3). Enterprise management behaviour was determined by the need to fulfil physical output targets; funds and inputs were largely supplied by the State, which also controlled the distribution of output from the predominant state-run sector. Such a system resulted in high risk aversion towards adopting new technology, few incentives to produce new products without guarantees of access to more resources, and few pressures to reduce costs or improve products. This is not to say that the system produced no technological change. Technology development activities were however managed administratively with few supplier/end-user feedbacks, there was much fragmentation of effort between different industrial departments and regions and a lot of innovation activity at the enterprise level was directed at the repair and maintenance of existing equipment, the building of new machinery for a plant's own use, or developing backward linkages to reduce reliance on outside suppliers. A "cellular" industrial economy thus emerged, stressing maximum self-sufficiency at the enterprise, regional and sectoral levels, with little priority in practice put on specialisation, co-ordination and economies of scale. A further constraint on rapid technological development arose from the structure of scientific and technological resources, a significant proportion of which were located either in a self-contained defence system or in the formal research and development sector, both of which were largely separated administratively from the civilian production sector. The lack of linkages has led to very low rates of technology transfer between each of the systems. This has especially influenced the development of advanced technology sectors such as electronics, computers, new materials, telecommunications, avionics and new materials, as these have been located largely within the defence industry system, with little diffusion to the parallel civilian sectors. The imbalance is now being acutely felt as it has proved extremely difficult to dismantle the old structure and develop integrated military/civilian systems.

One cause of widespread technological stagnation has been the role played by the machine-building sector over the years. A modern machine-building capacity took shape in the 1950s through the large-scale import of complete plants from the Soviet Union and Eastern Europe. These large factories were rather dated by international standards even then but have remained the core of China's industrial system as they were gradually duplicated. The original equipment has gone through a number of major overhauls but has not been replaced on a large scale by improved equipment. The machinery made by newer plants has not been significantly improved, either, so that while machine-building capacity has registered impressive growth, this has often been on the basis of "the replication of antiques" (4). Much of the machinery produced in the 1960s and early 1970s has long passed its useful service life, which because of very low depreciation rates is often set at twenty to twenty-five years. Large resources have been invested in its maintenance and repair rather than in designing and producing new or improved models. Some indications of technological backwardness can be inferred from the following indicators. One quarter of all industrial fixed assets are operating beyond their set service life. In 1988 new products accounted for just 7 per cent of the industrial output value of China's 10 000 largest enterprises (5) (which themselves produced over 40 per cent of national industrial output value). In the machine-building industry only 11.5 per cent of all its products by value were classed as national or ministerial first grade. A 1985 national survey of the levels of production equipment of key factories estimated that 13 per cent was of international standards, 22 per cent domestic advanced levels, 47 per cent were ordinary and 18 per cent backwards. Of the 8 million workers in large and medium enterprises: 25 per cent worked with machinery and 34 per cent with machinery some of the time, while 39 per cent worked by hand (6).

Numerous studies of regional and sectoral technological levels indicate considerable differences in levels, but it is clear that other factors, such as quality of management, have a significant impact on productivity. For example, in the Shanghai light industrial sector over two-thirds of the equipment is of 1940s and 1950s design with only 9 per cent of 1970s level, but productivity is still higher than in many other cities which are better equipped (7). The recent, albeit crude, measures of the contribution of "technical progress" to industrial growth in a number of major industrial bases during the 1981-85 period shows the wide differences between cities with Shanghai, Beijing and Tianjin having 35, 30 and 19 per cent respectively. Little store can be put on such figures themselves, but they do indicate the existence of unexpectedly wide variations between essentially similar regions in their response to using technical change as an instrument for economic development.

Faced with years of slow and patchy technological change and ambitious macroeconomic objectives, Chinese policy-makers have embarked on formulating and implementing a reform package which includes both policies and measures designed to improve the domestic supply of industrial technology and to stimulate demand from the industrial sector. In addition, policies to acquire and assimilate foreign technology have been implemented, though measures to co-ordinate these with domestic technology development have proved difficult to formulate, with repercussions for both sets of policies. The following section examines the first set of policies, those concerned with domestic supply and demand.

Domestic supply of new industrial technology

The structure of the S&T system and allocation of S&T resources in China was initially based on the Soviet model, and despite a number of experiments to modify it, the system has largely remained intact over the years, though most recent reforms have led to potentially significant readjustments and a break with the traditional system. The overall contours of the S&T system are as follows. A significant proportion of human, financial and material resources are concentrated in the formal S&T sector, not in the production sector. This is especially true for R&D resources. The formal sector itself is composed of five main actors, the Academy of Sciences, R&D units subordinate to central ministries, those run by local governments, the higher education sector and the defence sector. A de facto division of labour operates between these actors with the Academy and the higher education sectors concentrating more on fundamental and applied research, the ministry and local sectors on developmental work, while the defence sector, still largely self-contained, carries out all types of R&D and trial-manufacturing work. In practice, there has been a blurring of responsibilities, as economic pressures force all R&D actors to participate in activities more directly related to immediate production needs because of reduced direct state allocations to R&D units. This process, however, has as yet had less effect on the ministry and Academy sectors, whose activities are still largely controlled through the central planning process. In the increasingly important area of advanced technology development, the major resources are mainly concentrated in the defence sector, with some spillover in the Academy and industrial ministry sectors, as some components of the former revert to civilian status. The degree to which the recent structural reforms in the defence sector have improved the transfer of technology to the civilian sector or stimulated the creation of an integrated military/civilian industrial complex is questionable, however. The available evidence implies that the considerable S&T resources controlled by the sector, especially in

such fields as computers, control systems, avionics, new materials
and electronics, are still largely directed towards military-related
objectives, with relatively little diffusion to related civilian
areas.

Within the formal civilian R&D sector, the key resources are
concentrated in the ministry and local sectors, with the former
controlling 54 per cent of scientists and engineers and 58 per cent
of total available funds. (These figures do not include the
resources in the higher education or defence sectors) Comparisons
with the S&T resources in the industrial production sector are
extremely difficult to make due to definitional problems. However,
the large-scale production sector has 56 per cent of the number of
scientists and engineers employed in the R&D sector, who work on
technology development activities, but has 77 per cent of the
latter's expenditure. The relatively greater expenditure is
consistent with the nature of the work done, which involves costly
product/process development, pilot and trial-production, etc.

In absolute terms the total resources allocated to S&T work are
large but in relative terms they are rather modest, with a falling
proportion of state expenditure as a percentage of national revenue
and around 1 per cent of GNP devoted to S&T acitivites. These crude
measures of inputs, however, tell only part of the story as the
important question is, how effectively are available resources used,
and what impact has this investment had on China's economic growth
and development. The measurement of "output" is a very difficult
procedure, and commonly used surrogate indicators such as technology
flows and patent statistics cannot be used for China at present.
The few statistics available, however, strongly indicate that until
recently the formal R&D sector contributed little to industrial
development in terms of the supply of new or improved products,
processes and services. Transfer rates of R&D results to production
ranged from 10-30 per cent, depending on the research performer.
Since the renewed emphasis on work that is more relevant to
production needs, the transfer rate is claimed to have risen
dramatically. While this may in fact be the case, the actual

economic impact resulting from increased rates of transfer may have been much less in practice than the aggregate figures suggest.

This situation arises from a number of factors. The first is that many R&D projects are now of very short duration, aimed at solving minor production problems met by very small production units. Secondly, "successful" transfer from research to production is in practice often measured in technical, not economic terms. This is especially true for the larger transfer projects, which may or may not achieve satisfactory economic results, as they are often funded by direct grants, and a range of subsidies are given for their trial-production and production. Thus decisions on choice of technologies are sometimes taken on technological or strategic grounds rather than on economic ones, as in a distorted pricing system where many investment decisions are made by administrators, economic profitability can be manipulated to fit the technology. From the enterprise viewpoint, if technology is transferred and the new product/process proves unprofitable, it may not be discontinued as the enterprise can bargain for subsidies, price rises, low cost inputs, tax breaks, etc. The "soft budget constraint" in which many state-run enterprises operate, especially the large and medium ones, has therefore in some respects a distorting effect on the domestic technology transfer process. Finally, there is considerable evidence that much industrial technology which is transferred and incorporated in new products is only exploited by a few enterprises and is rarely diffused widely to other factories making similar products.

Diffusion is a topic which is rarely discussed in China's specialised economic and technology policy press, a surprising omission given the many instruments available to the state to promote rapid dissemination through control of the R&D, investment and production processes and the fact that an important factor in maximising economic benefits of investment-in-technology development is the broad dissemination of best-practice techniques. Scattered evidence confirms the low rate of diffusion. In a large-scale survey of the transfer of 144 R&D results to production units in 1986, over 60 per cent were implemented in one enterprise only,

17 per cent were taken up by two enterprises, and only 16 per cent by five or more (8). Examination of the spread of major process innovations in relatively homogeneous product sectors such as steel and cement shows a similarly slow adoption among large producers (9). The low rate and speed of diffusion is a function of many influences. The pressures on enterprises to be as self-sufficient as possible and their long tradition of keeping equipment in service as long as possible have resulted in many large enterprises routinely making their own replacement or new additional equipment, parts components. A 1985 survey of 1,800 electrical equipment and machinery plants found that by value, only 42 per cent of the parts and components used in their products were procured from other enterprises (10). This proportion dropped to around 15 per cent for the production of specialised equipment such as metal processing machinery. Such high rates of self-supply in crucial sectors point to low levels of intra- and intersectoral technology transfer, inefficient production due to low economies of scale, and significant constraints on efficient diffusion, as much capital equipment is produced on a customised rather than a serialised basis, with low standardisation and specialisation. Such a situation contrasts sharply with that in more industrially advanced countries, whose industrial system exhibits close and symbiotic networks of intra- and intersectoral linkages, leading to rapid flows of technology and its dissemination (11).

The poor performance in generating new technology on the one hand and applying it effectively on the other hand can be explained by a wide range of factors and constraints. A number of measures have been introduced since 1979 both on the supply and demand sides to improve performance, with varying success. The following briefly examines some of the major reforms being implemented on the supply side. A major reassessment in S&T policy in the early 1980s established guiding principles which are still operational today. The intervening period has witnessed the formulation, implementation and, where necessary, the modification of measures to put these principles into operation. The main thrust was to apply available S&T resources to promote rapid economic and especially industrial development. The key features were: to improve the translation of

R&D results into "direct productive forces", i.e. to stimulate technological innovation (12); to facilitate the transfer of technology from advanced to backward regions and from the military to the civil sector; gradually to expand resources devoted to basic research; to gain access to technology from abroad and assimilate it in order to improve domestic technological capabilities. In practice, measures have concentrated on reforms of the R&D system in the areas of research/production linkages, financing R&D activities, introducing greater automomy at the institute level, and a range of personnel- related measures aimed at more effective use of human resources. The reforms have been applied cautiously, often in the face of resistance, and have achieved varying success.

Chinese commentators have identified the lack of appropriate linkages between research performers in the formal R&D sector and production enterprises as a major cause of the small flow of technology from the former to the latter. The composition of the industrial sector itself complicates this issue. Around 10,000 large and medium plants make up the core of Chinese industry, accounting for over 40 per cent of total output value. The remaining output value is produced by around 400,000 small state and collectively-owned enterprises, with an increasing proportion accounted for by rural industry. The latter two categories contain only a small proportion of the production sector's total technical and engineering staff, and their technical levels are in general far inferior to those of the large and medium plants. They face the dilemma of an acute need to gain access to new technology while having little formal channels to do so and few resources to upgrade their equipment or assimilate any technology acquired. The situation of the large and medium sector is rather different, as most technical and engineering resources are concentrated in these plants. The volume of these resources devoted to technology development activities is relatively low, however, and there are many complaints of their underutilisation. In addition, there is a highly skewed distribution of S&T resources between different sectors, with the major part concentrated in the heavy industrial sector. There has been a dramatic growth in the number of R&D units or departments established by large and medium enterprises in the past few years, and a concurrent shift of resources to technology development activities (13). However, the number of these set up jointly with independent R&D institutes from the formal sector (3 per cent) indicates the lack of integration of the resources of the two sectors, despite the administrative pressures applied since 1987 to promote the merging of research and production units.

While the vast majority of technology development departments have been established by large and medium enterprises using their own internal resources, a wide range of looser forms of co-operation have been initiated with units in the formal R&D sector. Statistics indicate that by the end of 1987, over half the R&D units in the latter sector had established some type of collaboration with production enterprises. The forms included mergers, R&D units taking over small and medium enterprises to commercialise their research results, institutes becoming sectoral or regional technology development centres, or joining up with design and production units to form engineering companies. The most common form of collaboration, however, has been the ad hoc joint development of specific technologies or products with one or more production enterprises. Overall, however, it is probable that for many larger enterprises, the buying of technology from the formal R&D sector or joint collaboration in developing new technology or services is not their major source of new technology. Scattered evidence indicates that these enterprises either develop their own technology themselves, prefer to buy it from other production units, or try to import it. This is because more appropriate and complementary technology is acquired through these channels than through research institutes and universities, whose results often need a lot more investment and work before they can be applied in production (14). Several large-scale surveys confirm such trends. For example, one investigation found that two-thirds of technology acquired by its sample factories was either self-developed or imported from abroad (15).

The more effective use of S&T resources through strengthening research/production linkages has been intimately bound up with a basic attitudinal change over the nature of technology. Until the early 1980s it was regarded as a "free good" developed largely through state funding and transferred and disseminated gratis through the industrial system. The switch to a "planned commodity economy", the introduction of limited market mechanisms and the encouragement of competition between enterprises, however, laid the basis for treating technology as a commodity which can be traded on the market and which has a potential value. This has had a positive effect on the R&D system, as it now has much greater financial incentives to commercialise its R&D results, although it is also under great pressure to do so due to the phased withdrawal of direct government grants. The total volume of technology transactions, however, although it has grown extremely rapidly from 50 million yuan in 1983 to 3.4 billion yuan in 1987, is still extremely modest, accounting for just 0.6 per cent of the total commodity trade, and very few units in the formal R&D sector are anywhere near self-financing through supplying technical services, selling technology or through R&D contracting, the three major forms of earning income (16). Indeed, it is clear that for most technology- oriented R&D units, a major source of income outside of that traditionally supplied by the state now comes not through the above three channels, but through the sale of components and equipment, i.e. technology embodied in equipment. In total, R&D units only accounted for 40 per cent of total technology contract value in 1987. A significant proportion of R&D projects undertaken by the formal R&D sector are thus still funded directly or indirectly by the state through the plan (central, local, sectoral, etc.) and the decision to transfer those that can be used in production is made mainly through administrative procedures, although a proportion of planned projects are now transacted on the technology market.

While closer collaboration between the formal research and production sectors has been a major platform of the process of improving transfer and assimilation procedures, a number of major constraints and barriers still exist to hinder these activities, thus reducing the rate of industrial technological development and the improvement of domestic technological capabilities. Past major successes have tended to be a function of high level administrative intervention or related to military programmes, a situation which could not be applied to non-priority transfer and dissemination projects, which are the bread and butter of industrial technical change activities. The main problems identified are as follows. Diffusion is constrained by a number of factors. Regional rivalries are important as it has proved difficult to disseminate technology across administrative boundaries for fear that competitive advantages are eroded. This helps to explain the extremely wide differentials in productivity between similar enterprises in different regions. The response of these "technical blockades" has been for regions to require their own enterprises to source as many products within their own regions and also to subsidise their own inefficient enterprises. Individual regions also often require R&D work they need to be done locally even though the expertise and resources may be inferior to those of other regions. The consequences of regionalism for technical change have been and continue to be severe, as transfers are limited, diffusion is slowed down, R&D resources are fragmented and projects duplicated. The profusion of small sub-optimal R&D units is a further outcome. Much of the same criticism can be levelled at rivalries between industrial departments, as the production of similar products is often carried out by a number of different ministries. Attempts to co-ordinate R&D and production activities across regions and departments for important products like electronics through rationalisation and the establishment of special co-ordination groups have only met with limited success, as the parties concerned have clear vested interests in retaining control over scarce resources, and the technology market is, as yet, an imperfect transfer and diffusion mechanism.

At the micro-level, a number of specific constraints operate to hinder more effective integration of R&D and production activities. The price paid for the transfer or contract work has been a source of friction between research and production units, as no generally accepted rules have been developed. Enterprises which have traditionally been used to receiving new technology without having to pay for it are now loath to accept that this is no longer a free good, especially since it often has to be paid for from bank loans or enterprise profits. Research units, with some justification, complain that transfer fees, etc., often barely cover their R&D costs. This has a number of consequences. In order to increase income to substitute for falling government allocations, R&D units are forced increasingly to take on short-term, "low technology" projects. While this is acceptable for small, local units, it probably signifies a misallocation of resources when applied to centrally run institutes. Secondly, many research institutes are either being forced into commercialising their own technology because it cannot be sold to production units, or do so as this can increase their income substantially, much more than pure transfer or research contract activities.

Such a trend has a number of positive and negative outcomes. The former include the manufacture of specialised products which would otherwise not be available or which would have to be imported. The trend has also resulted in the establishment of a large number of small, specialised high technology firms. Such start-ups, common in industrially advanced countries, have until recently been missing from the Chinese industrial map. Their rapid profusion, especially as spin-offs from such high level institutes as the Academy of Sciences, indicates both the difficulty of transferring research results through normal channels and also the growing demand for products incorporating advanced technolgy, a demand which is not being met by large-scale plants in such sectors as electronics, materials, etc. On the negative side, however, many R&D units are ill-equipped to start manufacturing operations, and their scarce resources are diverted from R&D activities. In the case of new spin-off companies, they are often greatly undercapitalised and have difficulty in procuring the necessary special equipment parts and components for their new products, which often have to be imported. They have also sometimes had a hostile reaction from those parts of the industrial system traditionally involved in producing similar types of products.

The policy of encouraging research institutes and production units to enter into various types of closer, long-term collaboration also faces a number of problems. The main one is that neither side is very enthusiastic about such arrangements. A number of cases exist of enterprise-run R&D units trying to break away and become independent, often because they have little work to do due to the lack of demand for new technology from the enterprise concerned. A number of surveys have concluded that most research institutes either cannot find a suitable partner, fearing that close collaboration would adversely affect their own development, or that such linkages are simply not in their own interests.

Some of the implications for the technology transfer and dissemination process of reforms to the funding of R&D have already been touched upon. Their overall thrust has been to increase incentives, diversify sources of investment for R&D projects, lighten the burden of the state, and introduce greater responsibility in selecting R&D priorities which correspond more closely to production needs. The corresponding development of a technology market has also created an initial framework through which more technology can be commercialised. The overall consensus is that these two reforms have had a beneficial effect on the supply of new technology for industrial development. A number of factors have, however, limited this effect and the direct economic benefits accruing. In 1987, almost 60 per cent of total income of the local and ministry-run R&D sectors and the Academy of Sciences still came from government grants, 32 per cent from non-government earnings (of which two-thirds came from technology-related income such as contract work, technology transfer, consultancy and trial production activities, and one-third from non-technology-related income) while only 9 per cent was obtained through bank loans and other sources (17). While it is intended to eliminate state funded operating expenses of technology development institutes by 1990, the increasing difficulties met by many such institutes in making up the loss of direct grants is placing clear limits on the pace and scale of commercialisation of technology development activities. Pressure to generate mere income has led to a tendency to ignore state assigned projects in favour of outside work; some research units are becoming de facto production units, and there is some evidence that basic research is being neglected. A further problem is that those units most successful in generating outside funds are the ones whose "safety net" of state grants is reduced the fastest, thus penalising success whilst poorly performing institutes continue to receive subsidies and grants. Under such conditions, enforcing financial discipline and tying access to resources more closely to performance are difficult tasks to achieve. Although the above problems are constraining factors, the financial reforms have only been implemented a short time and scattered evidence does show that they have had an effect on the behaviour of some institutes. The selection of new R&D projects, for example, is now done more with the objective of meeting production needs.

There is a wide consensus that the movement of technical, engineering and scientific personnel is a critical factor in the technology transfer and dissemination process. The measures implemented in China in recent years concerning personnel-related issues are in fact a response to this consensus. The issue has several dimensions. Extensive misallocation of S&T personnel has resulted in a great waste of what is a scarce resource. An even bigger problem is the widespread underutilisation of talent, with S&T expertise locked up in institutes and enterprises which do not exploit this expertise, while at the same time other organisations cannot get access to expertise which they desperately need. Attempts to tackle these problems through loosening the extremely tight control exercised over mobility of the S&T work-force have only had a very limited success to date. Apart from the strong vested interests exhibited by units in retaining control over such a scarce resource even though it could be more effectively deployed elsewhere, the question of greater mobility raises complex political and ideological issues, given the ambiguous position of intellectuals in Chinese society. The extremely low mobility of personnel within the formal R&D sector, between the research and production sectors, between enterprises, and geographically, is seen as one of the major constraints on improved transfer and dissemination activities, given that technology cannot be completely embodied in products or codified in documents.

Statistics for S&T mobility in fact show declining trends in the past few years from 2.7 per cent of total S&T personnel in 1984 to 1.9 per cent in 1986. Such low levels clearly greatly inhibit flows of knowledge and expertise throughout the system. One response to difficulties of outright transfer has been to allow secondment, holding concurrent posts, leave without pay, or after-hours work by S&T personnel. The latter especially has been of great help to small urban and rural enterprises who have little or no access to such expertise through traditional channels. These forms of mobility have also been important in allowing S&T personnel to establish non-government technology or product development companies, consultancy and technical service activities etc., i.e. to encourage the formation of a new economic group of "scientist-entrepreneurs". Such activities do appear to fill an important need which is currently not being serviced by either the R&D or production sector, especially in the start-up of small high-technology firms. However, these entrepreneurial efforts are taking place in a generally hostile environment; the new companies are often starved of resources and cannot easily gain access to a range of inputs. They cannot therefore be compared to small entrepreneurial firms which play such an important role in the innovation and diffusion process in certain economic sectors in industrially advanced economies.

On balance, policies and measures designed to improve the supply of new technology have succeeded in developing incentives and applying pressure to R&D units to re-orient their efforts more towards the needs of the production sector. The process is still thought in the early stages, however, as the R&D sector is subject to continuing administrative intervention in matters concerning the allocation and use of its resources. The extensive commercialisation of technology is a long way off as vested interests, structural rigidities and economic factors combine to form a less than optimal environment for the reforms to operate in. Research/production linkages are still often rather tenuous, and while the reforms of the system of funding R&D activities have introduced an element of flexibility which allows certain parts of the R&D system to respond more to production needs, this flexibility is constrained by a number of factors. This is reflected in the technology market, which is currently relatively small and highly imperfect. Transfer and diffusion is also constrained by the low mobility of the scientific and engineering work-force, whose entrepreneurial potential is heavily circumscribed. The development of innovation channels which circumvent the inefficiencies and rigidities of the current system are therefore limited. While therefore the supply of new technology has improved (though not as fast as was expected), the key determinant in the Chinese case appears to be the volume and type of demand generated in the industrial production sector and the factors determining effective demand.

The demand for domestic industrial technology

Given the comprehensive range, composition and geographical spread of industry in China, it would be expected that demand conditions vary widely. The 9,000-odd large key enterprises, 100,000-odd small state-run enterprises, 1.8 million collective rural and urban enterprises and 5.6 million private rural units clearly have different needs in respect to access to, and use of, new technology. The economic and especially industrially related reforms have also had markedly different impacts on demand patterns, and the S&T reforms discussed above have, in part at least, responded to this differentiated demand.

At the broadest level, there is a wide consensus in China that the past situation of sluggish demand for new technology has not altered significantly despite the economic and industrial reforms initiated in the past ten years. A number of factors are in play which dampen the demand that the authorities are trying to stimulate as a key facet of their overall industrial development policy. The first is attitudinal, as although an intensive growth strategy has been official policy for many years, many economic planners, personnel in charge of administering policy and factory managers themselves are still actually pursuing extensive growth strategies which traditionally have placed little emphasis on growth through rapid technological change. A rigid price structure continues to distort innovation patterns, because as long as many enterprises continue to obtain inputs at prices which do not reflect their scarcity or cost of production, they face few penalties or have few incentives to introduce cost reducing or input saving innovations. Poor demand is also a function of the high risks and costs incurred by enterprises which innovate or adopt new technology but may not reap any economic benefits from doing so.

The intense pressures on enterprise management for short-term increased output, pressure which comes both from above and from the enterprise work-force, is a major factor in dampening enthusiasm for technological change, which can be complex, costly and time-consuming. Inadequate sources of funds is also a commonly cited reason for the slow pace of industrial technological change. Commercialising new technology often requires many times the investment in the initial R&D work. Financial reforms have increased the number of channels for funding such work, as the emphasis has shifted mainly from state grants to a combination of grants, special loans and a proportion of enterprise profits. For the large scale industrial sector, the proportion of total technology development funds from the above three channels in 1987 was 11, 40 and 45 per cent respectively (18).

The degree of self-funding is rather high and probably reflects the special position of the large-scale sectors, as the common complaint by enterprises is that after payment of different taxes, most only retain 10-20 per cent of their gross profits, much of which is diverted to payment of bonuses and subsidies to the work-force rather than on production development projects. Enterprise technology development funds average out at between 0.2-0.3 per cent of total annual sales (19). This average increases to around 1.5 per cent for China's largest enterprises but many medium and most small enterprises have no such funds at all. This very low rate of investment in development activities is boosted by other sources of funds for re-equipment activities, technology imports, etc. It is interesting to note that funds for import for China's 9,000-odd largest plants were virtually equal to those for technology development activities in 1987, while re-equipment funds were 2.5 times (20). The funds available for assimilation and purchase of domestically generated technology are very limited indeed, and the low rate of funding such activities as new product development must be a cause of great concern to the authorities, given the current rate of technological change outside China.

Available evidence suggests that the above constraints do not operate uniformly across industrial sectors, nor do they affect enterprises of different sizes in the same way. For example, high priority and advanced technology sectors such as electronics tend to have more R&D funds than mature sectors. Large-scale sectoral surveys have shown that technology development funds as a percentage of sectoral output value vary widely from a low of 0.5 per cent in the metallurgical sector to 1 per cent in the machine-building and chemical sectors to a high of 5.5 per cent in the electronics sector (21). In terms of openness to technological change, large and medium-sized enterprises tend to be the least dynamic while the collective sector, which is not cushioned by preferential access to resources, is the most enthusiastic. Strategically, because of their role in industry in terms of output value and key producer products, changing the behaviour of large enterprises towards technological change is perhaps the most important and complex task facing Chinese industrial policy-makers and planners, a view reinforced by the spate of reports on this subject in the past few years.

A number of direct and indirect measures have been introduced in recent years to stimulate innovation by industrial enterprises and to encourage them to adopt new technology developed elsewhere. These range from efforts to increase the technological capabilities of individual units through close links or mergers with R&D units to financial incentives to make the production of new and improved goods more attractive. Other measures include allowing higher prices for better quality goods, allowing enterprises to recoup part of their development costs, and a more liberal tax policy for innovating enterprises. Pressure is applied by progressive taxation on obsolete products, raising material and energy prices to induce cost-reduction innovations and reducing supplies of low-cost state allocated inputs. In addition, proposals have been made to tie management and work-force incomes more closely to technological improvements achieved rather than to gross output or profits, in order to provide enterprises with incentives to adopt new technology. This wide range of instruments has only been applied very recently, and it is unclear as yet what effect they have had overall or individually, by sector or by size and type of enterprise.

In more general terms, the industrial reform measures introduced since 1984 have aimed at increasing enterprise vitality through increasing efficiency, inducing innovation and promoting intensive growth (22). One key has been the expansion of enterprise autonomy in a number of areas. The degree of autonomy achieved varies according to sectoral and size characteristics, with enterprises in capital goods industries and large, key enterprises gaining the least autonomy. Investment and pricing autonomy are clearly related to an individual enterprise's control over its own technological development. Greater autonomy is seen as a precondition to encourage technological change but there is strong evidence that the greater funds for technological development now under the control of enterprises (for example retained profits, access to bank loans) are often diverted to consumption or capital construction. Partial investment autonomy therefore does not necessarily promote innovation to product improvement. Greater autonomy to set product prices, in an environment of incomplete information and highly imperfect markets, does not necessarily encourage competition or, by extension, the use of technology as a means to improve competitiveness. In a situation where for the vast majority of products demand exceeds supply, price competition through improved productivity, etc., is difficult to promote. Greater autonomy has not been tied closely to greater responsibility for profits and losses, and enterprises are often not accountable for their losses because of the soft budget environment in which many operate.

Enterprises can only become independent commodity producers if they are responsible for their own economic results and operate in a market where there is free circulation of funds, labour, material inputs and products. A wide range of sectoral, regional and administrative constraints has hindered reforms in all these areas, and protectionism largely provides a barrier behind which inefficient and backward enterprises can continue to operate, as exit pressures are very weak while entry barriers are also low. The low degree of autonomy granted to large key enterprises is now regarded as the key issue by policy-makers and explains in part at least why the industrial reforms have yet to increase their vitality.

The reforms of planning and supply of inputs have altered enterprise behaviour, but it is difficult to trace their effect through to the latter's propensity to adopt new technology. The proposed inclusion of technological targets in measuring plan performance could have some effect on behaviour, but as plans are neither taut nor firm, they tend to have little effect on enterprise incentives, as individual bargaining on targets is widespread. The supply system is characterised by multi-channel allocation and soft supply constraints, and the reforms instituted in this area have mixed implications for stimulating innovation. Poor quality and lack of variety of low-priced allocated inputs may force enterprises to acquire inputs outside the plan despite the higher costs, which may induce them to help improve their suppliers' operations. Inefficient producers are, however, protected in the current system, and the strong local administrative pressures on enterprises to source their inputs locally adds to this. Local maximum self-sufficiency appears in many respects to be as strong as ever despite reforms aimed at rationalisation, and this tends to perpetuate inefficient producers. Indeed, the quality of industrial products has declined in some respects in recent years.

The increasing role of market mechanisms has had an important effect on the behaviour of some enterprises in recent years, depending on whether the enterprise faces excess demand or supply for its products. The response to excess demand can range from doing nothing to increasing production to meet demand, although this does not usually include attempts to cut costs, improve quality or introduce new products to capture greater market shares. A sellers' market therefore is a significant constraint on technological development. Excess supply has tended to induce a number of more complex responses. Initial appeals for help from local authorities may be followed by more vigorous sales promotion. This may shift to more attempts to meet customer needs by changing product mix, raising quality and improving products. Attempts to reduce costs, however, are more rare, as the burden of lower revenues may be shifted to the local authorities, while exit from production of goods in excess supply is an uncommon response. The response to reduced production planning and distribution controls in favour of market mechanisms is therefore conditioned by a number of factors, which can vary widely between enterprises and regions. In general, however, the macroeconomic reforms have not succeeded in producing an environment conducive to generating a strong demand for new technology by industrial production units, although there have been some changes in behaviour towards developing or adopting technology.

The role of foreign technology in industrialisation strategy

Since 1949 attitudes towards the place and role of foreign technology have fluctuated wildly in response to shifts in domestic development strategy, the prevailing ideological line and the changing external environment (23). It is no exageration to say that China's modern industrial base was developed by the massive flows of technology, equipment and expertise from the Soviet Union and East Europe in the 1950s. Over 150 projects were completed in this period, located in all parts of China but concentrated in the traditional heavy industrial bases in the north-east. Transfer mainly took the form of turnkey plants and complete sets of equipment. This was accompanied by the provision of large amounts of technical data and blueprints, as well as intensive training. The programme was very successful both in terms of the rapid development of key production capacity, mostly in heavy industrial sectors, as well as the acquisition of technology development and design capabilities. This co-operation, however, came to an abrupt halt in 1960 with the Sino-Soviet ideological rift. China's major source of modern technology and equipment disappeared overnight and a number of half-built projects were halted, only to be finished

years later. This bitter blow was compounded by the ill-conceived
industrial policies pursued in the late 1950s, and it was not until
the mid-1960s that output of many industrial products regained the
levels of the mid-1950s.

During the 1960s China secured new suppliers in Western Europe and
Japan, but the flows were drastically reduced to one-tenth that of
the 1950s. Complete sets of equipment predominated but a number of
transfer problems resulted in poor rates of return on this phase of
investment. The Cultural Revolution intervened to delay the
completion of a number of projects, and flows of equipment and
technology from abroad virtually stopped in the late 1960s and early
1970s. By 1973 China was poised once more to restart large-scale
imports of turnkey plants. Over $3 billion was spent on 26 large
industrial complexes, with emphasis on petrochemicals. A number of
serious problems dogged the implementation of this programme and
caused major losses. In many cases the lack of feasibility studies
led to extensive co-ordination problems in the supply of inputs,
construction schedules were not met, the siting of many projects was
irrational, the domestic supply of complementary equipment was often
held up and returns on investment were very poor. A survey of
30 large turnkey projects built in the 1970s concluded that only
one-third were considered satisfactory, measured by their
construction times, post-commission utilisation rates and
operational results. Virtually all the projects overran their
construction schedules, eleven by over a year and some by over three
years. Of the 17 projects completed by 1978, only nine reached
90 per cent of their designed capacity while the six largest
projects, which accounted for half the foreign exchange costs of all
projects, reached less than 50 per cent capacity (24). In some
cases, the claims that good economic results were achieved had more
to do with manipulation of input and output prices rather than with
efficient operation and utilisation. The concentration on importing
complete plants, often with a multiple import of similar complexes,
is now seen as a serious policy mistake. Extreme difficulties were
faced in unpackaging the technology embedded in these plants, partly
because of the large technology gap between the supplier and user.
This led to extremely slow accumulation of technological

capabilities and only a gradual move from production to investment and finally independent innovation capabilities. For example, after the repeated import of large chemical fertilizer complexes since the mid-1970s, capabilities have so far only reached the investment stage. Part of the problems met in this phase is blamed on the overcentralisation of project planning and implementation, as the most advanced technology available tended to be selected and there were few mechanisms to keep costs in check, as they were supplied by the state in the form of direct allocations.

Few lessons were learnt from the above experience, and another ambitious import plan was drawn up and partly implemented in 1978, following the formulation of an eight-year economic modernisation plan. Many of the contracts signed were subsequently modified, postponed or cancelled, as it quickly became clear that the economy could not sustain the rapid growth envisaged. The resulting reassessment of economic strategy in the early 1980s included a reappraisal of the role of technology and equipment imports in future development. Technology flows declined rapidly after their peak of 1978-79, only starting to rise equally rapidly from 1983 onwards. Data on total flows between 1983-85 are ambiguous as there was no unified definition of what constituted "technology imports" until 1985, and even after that, different departments used different ways to measure flows. The commonly used figure for this period is $10,billion, involving 13,000 contracts. Of this, 3,000 projects costing almost $4 billion were implemented under a special technology renovation programme.

The new approach emerging in the early 1980s included some significant shifts from previous policy. Emphasis on a small number of large complexes was discarded in favour of the import of key equipment and technology in the form of licensing, consultancy, etc. The size of contracts was much smaller, as strategy focussed on upgrading the technical level of a large number of small and medium plants. Certain industrial sectors, including the electronics, light and machine-building industries, were given investment priority. The greater autonomy granted to certain cities and regions on the eastern seaboard in foreign exchange retention and

use introduced a de facto spatial bias into technology import
strategy. Concurrent changes were also made in breaking the
monopoly of central trade agencies handling technology import work.
These trends have been largely carried over into the current phase
of the Seventh Five-Year Plan (1986-90), although greater emphasis
has recently been placed on improving the technical levels of large
and medium plants, and technology flows continue to fluctuate from
year to year, reflecting the regular imposition of import controls
as an instrument of macroeconomic control. The target set in the
past few years of allocating 10 per cent of total export earnings to
purchasing technology was not met in 1987 or 1988, although no
explanations have been furnished for why this was the case.

An examination of the degree to which technology import practice has
followed the above policy shifts shows the following. Technology
flows peaked in 1986 at $4.5 billion from the very low levels of the
early 1980s. Despite the policy shift from complete sets to
individual pieces of key equipment, the former still formed the
largest proportion of imports up to the present. For example, in
the peak year of 1986 a small number of contracts for a nuclear
power station and several large thermal power stations accounted for
almost two-thirds of total spending. Of the "pure" technology
import contracts, licensing was by far the most important vehicle
for technology transfer, though the proportion by value of "pure"
technology transfers is still much lower than has been recommended.
In recent years, foreign loans have financed a significant
proportion of total contract value, and the major suppliers of the
1970s, i.e. Japan, the United States and the Federal Republic of
Germany, continued to be so in the 1980s, although some
diversification has taken place. Although many more agencies are
now involved in the technology import process, the traditional
central agencies are still major players. In addition, regulations
on contract examination and approval have been issued which provide
instruments for greater central intervention and control in the
technology import process. While on balance the new directions in
policy have been partly adhered to, implementation has not been
uniform or total. The following sections examine in more detail
important elements of policy as it has evolved during the 1980s.

Technology imports and intensive growth strategy

Investment in improving existing production facilities as opposed to the construction and equipping of new facilities has been a major component of industrial development strategy in the 1980s. Funding of technological renovation has steadily increased, and while the upgrading of obsolete production facilities and the introduction of new products and processes has been achieved largely through domestic procurement and development, imported technology and equipment has played a growing role in this programme. Two different approaches can be identified. The first has been the initiation of a small set of projects to trial-produce and manufacture key sets of equipment which had previously been imported, often repeatedly. These projects, many of which involve extensive import, co-production and assimilation activities, have been given very high-level backing and priority and include developing equipment for nuclear power stations, civilian aircraft manufacture, large power generators and transmission equipment, chemical equipment, heavy locomotives, etc. Reports indicate varying success after five years work, although experience gained in the management of such large and complex projects should be of considerable use in the future, especially as import substitution activities are gaining in importance.

The second approach has formed a major component of technology impact work and has been largely directed at small and medium-sized enterprises. The plan for 3,000 technology renovation projects incorporating imported technology and equipment was drawn up by the (former) State Economic Commission in 1982 and implemented under a rolling plan between 1983-85 by the relevant industrial ministries and localities. The latter two authorities were given the power to examine and approve small projects, with the centre approving larger projects. In the end 3,900 import contracts were signed for a total of $3.6 billion. In addition various localities and ministries signed another 10,000-odd contracts. In practice, many of these were probably simply equipment or production line import contracts with little import of technology per se. Little information on the results of this large programme has been released, though scattered evidence indicates that the direct or indirect recoupment of their foreign exchange costs averaged around 3.5 years. In 1987 the department of the State Economic Commission in charge of the overall technical renovation programme estimated that 60 per cent of the total increase in national industrial output value in the mid-1980 came from factories retooled with foreign equipment and also from improvements in enterprise management (25). The approach to technical renovation for the 1986-90 period is to place more emphasis on large and medium enterprises. Geographically, investment is to be focussed on coastal regions to develop high-grade consumer goods for both the domestic and foreign market, as well as building up new advanced technology sectors. This emphasis is in line with the overall industrial development policy, which favours the rapid expansion of the eastern seaboard.

Spatial issues

Current policy has formalised a de facto regional division of labour, and technology development policy reflects this division. Coastal cities are to act as major conduits for foreign technology, master and assimilate it and then diffuse it to other regions. The process started with the establishment of four special economic zones in Guangdong and Fujian. The acquisition of advanced foreign technology through large foreign direct investment flows was a major objective in setting up these zones. In 1983 Shanghai and Tianjin were given special powers to manage their import work, and by 1985 one-third of technology import funds flowed into these two cities. Initially both focussed on upgrading their export industries, although as the number of projects rapidly increased, more investment and imported technology went into their advanced technology sectors. The impact of these flows on their local economies is difficult to measure with any accuracy, although some pointers can be given for Shanghai (26). Between 1983-87 the city imported over $1.1 billion of equipment and technology, excluding projects initiated by central industrial bureaux or local factories. In all, over one-third of Shanghai's key enterprises were retooled to varying degrees. Of the 500 new products trial-produced in 1988, it is claimed that almost 80 per cent relied on previously imported equipment and technology, which was also responsible for over half the city's increased industrial output value between 1985-87. While most reports on Shanghai's progress are positive, a few discuss less favourable factors. One such is the extremely small amount of funds allocated to assimilation activities, just one-fourtieth of that spent on technology imports. The relative success, however, is put down to a number of factors. Major ones include a strong existing industrial and technological base and especially the creation of special administrative agencies to oversee the whole import process, with special attention put on reducing bureaucratic red tape and organising the many different organisations involved in any one project to work together effectively. Despite these favourable features, the Shanghai authorities have pointed out that they consider their access to foreign exchange does not match their industrial strength. Fluctuations in foreign exchange availability also make long-term planning and co-ordination rather difficult.

While such cities as Shanghai and Tianjin have attracted many technology import projects, in terms of sheer volume of equipment, Guangdong province is by far the more important (27). This is because it is allowed to retain a large percentage of foreign exchange earnd, and it claims to have absorbed $5 billion of foreign investment, 40 per cent of China's total. Over 150,000 pieces of equipment and 2,000 complete assembly lines have either been bought or supplied by foreign investors, most of which is used to produce consumer and light industrial products for export. Although much of this increased industrial activity is labour-intensive assembly and processing work, the effect on the province's economy, especially the area around Guangzhou, has been dramatic. Proximity to Hong Kong has been the deciding factor in this process, combined with an aggressive export-oriented strategy and a willingness to put local interests above national ones where the two clash. The effect of this large inflow of investment and equipment on the province's technological capabilities is unclear. However, Guangdong has been criticised for not granting other units access to its imported equipment or to allow interprovincial assimilation activities to take place. Such technical blockages parallel those applied to domestic transfer activities and clearly both hinder diffusion and the improvement of domestic technological capabilities.

A further spatial bias was introduced in 1984 when 14 coastal cities were granted the same foreign exchange and technology import autonomy as the zones. This quickly proved to be over-optimistic as most of them did not have the conditions to utilise their autonomy, and in practice it was decided to focus on just four cities, Shanghai, Tianjin, Dalian and Guangzhou. By the end of 1987 the majority of foreign direct investment and investment in technology imports was concentrated in a small number of coastal cities and provinces. This trend dovetails into the recently adopted strategy, which sees the rapid economic and technological development of the coastal areas as the lynchpin of China's overall modernisation strategy. Since the fall of Zhao Ziyang, who was closely identified with this strategy, its implementation is in doubt. The current austerity measures aimed at cooling overheated economic growth, which were introduced in late 1988, have resulted in less foreign exchange being available to such cities as Shanghai, which has seen its funds for importing technology cut by half in 1989. Fast-growing provinces such as Guangdong have argued, with some justice, that they should be exempted from some of the current austerity measures, which provide for a recentralisation of previously delegated powers in a number of fields. While they have been partly successful so far in resisting some of the recessionary measures, by late 1989 even their economies were starting to feel the pinch. It remains an open question whether the centre's success in reining in the most dynamic growth poles of the whole economy will be outweighed by the damage done to longer-term confidence in the areas which have most closely started to integrate themselves into international markets.

Sectoral priorities

Given the wide technological gap with both industrially advanced
countries and the growing gap with some of its neighbours and
competitors, especially the Asian newly-industrialising economies,
China's industrial planners and policy-makers have faced an
unenviable set of choices in allocating scarce resources to upgrade
the country's industrial base. The restructuring programme of the
early 1980s aimed at redressing the imbalance between heavy and
light industry and building up the basic infrastructure. Imported
equipment has played a small but important part in certain aspects
of this restructuring process. In developing light industry to meet
years of pent-up unsatisfied domestic consumer demand and to expand
rapidly export earnings to provide the funds for further imports,
foreign equipment has played an important role in such areas as
textiles, consumer durables and electronics, whose output has risen
dramatically through the 1980s. Imported equipment has also been
instrumental in improving quality, variety and competitiveness in a
range of traditional consumer and light industrial exports. Imports
have been important in alleviating some infrastructural bottlenecks
such as power generation, and a significant proportion of China's
international and bilateral loans have been used for infrastructural
projects, although the role of foreign direct investment in this
area has been very disappointing.

The Sixth Five-Year Plan (1981-85) announced in late 1982 drew up a
list of priority areas for technology and equipment imports. The
electronics and machine-building sectors were singled out for
priority development, and there has been increasing preoccupation
since 1983 with the development of technologically advanced
industries, as policy-makers have become more aware of their
potential impact on overall economic development.

Data for the machine-building industry is incomplete, but
technology/equipment flows in the 1981-85 period have been
substantial, with conservative estimates of $1 billion. Many of the
contracts in this sector were licence, co-production and technical

service projects, and their impact is claimed to have been considerable, with 12 per cent of all product lines being improved (28). Since 1987 steps have been taken to try to reduce the rapidly growing volume of duplicated machinery imports by encouraging an import substitution programme through greater technology transfers, particularly through licensing agreements, which made up 50 per cent of technology contracts between 1986-87. The major aim of the technology transfer programme has been not so much to increase production capacity but to develop energy-saving products, complete sets of equipment, key basic components and higher quality goods for export. However, end users still prefer, wherever possible, to import machinery rather than buy similar locally produced products even where these have been developed with foreign technology. Measures to combat this trend include administrative controls on the import of equipment that can be produced domestically, allowing designated "import substitution" enterprises to receive part payment in foreign exchange for their products, etc.(29). The different interests between domestic producers and end users is a difficult one to solve in the short term, as undue protection of local producers can lead to a slow-down in the pace of technical change and the continuing protection of backward and inefficient producers.

The technology import policy pursued in the electronics sector, another high priority area, has had a somewhat different focus. The civilian electronics sector has been a traditionally weak area, as the majority of resources have been controlled by the defence sector, with very little transfer of technology between the two. The situation is complicated by the particular problems of access to sophisticated foreign technology in this sector, due to both external strategic and commercial considerations. Despite these barriers, between 1981-85 China spent over $1.4 billion on technology/equipment imports, and it is claimed that 30 per cent of its key electronics enterprises have been modernised. These imports have been responsible for the rapid development of the consumer electronics sector from a very low level. Product lines like tape recorders and TVs have grown exponentially, and China has become the world's largest B&W and third largest colour TV producer from being marginal producers just ten years ago. Investment electronics and components development has been much less dramatic, and the whole sector is considered to be unduly skewed towards the consumer electronics sub-sector.

Despite its priority status, there have been major problems in co-ordinating technology import, domestic technological development and production development measures. The policy dilemma is that in a rapidly growing and technologically changing sector, how does a country like China enter into such a sector? Both end-user needs and domestic producer markets have to be balanced, and the technological capabilities of the latter constantly upgraded. This usually entails large investments and initially high production costs until the technology is mastered and economies of scale are achieved. This is not helpful for the end-user, who can often procure better quality and lower priced inputs externally to keep own costs down and guarantee quality. Imports of electronic components and final products have been necessary for the rapid development of China's electronics industry, but have also hindered domestic technological and production development. The situation requires a very flexible and sensitive use of administrative regulations and economic instruments, a combination of approaches which China is currently ill-equipped to use. The response has been to impose protectionist measures to control certain categories of components and final goods, as well as to reorganise production facilities into large enterprise groups or corporations. Efforts have also been directed towards importing more know-how rather than equipment and assembly lines, and to try to implement unified plans to develop new sub-sectors such as video cassette recorders.

Increased emphasis has been put on the development of information technology as opposed to just electronics technology, as it is recognised that separate plans for electronics, computer and telecommunications miss the increasingly interlocking nature of these areas. However, a national technology policy on informatics was only formulated in 1988, and an integrated development strategy is probably still some way off. Chinese strategy in such crucial areas as informatics in theory recognises the importance of foreign

technology, but sectoral, structural and strategic interests combine
to frustrate a co-ordinated approach which would concentrate limited
resources and subordinate individual to national interests. The
formulation of a domestic high technology plan in 1986 and one to
commercialise these technologies in 1988 are pragmatic responses to
mobilise domestic resources and, through the latter plan, gain
access to foreign technologies to maximise the investment in the
former plan, though it is too early to analyse whether such
nationally-run programmes will overcome the fragmented approach to
technology development in advanced technology fields. As will be
examined below, however, technology import policy and planning in
certain consumer electronics areas has broken down completely, and
the impact has been felt throughout the sector. Similar situations
have occurred in many other less sensitive product areas, indicating
a degree of inability by the central authorities to control and
define development strategies of certain sectors, especially those
in which technology/equipment imports play a significant role.

Reforms of the administration and the management of the technology
import process

The major objectives of reforms in this area have been to simplify
and improve the efficiency with which foreign technology is acquired
and projects implemented, and also to ensure that the technology so
acquired is appropriate to China's specific conditions. The
devolution to local areas of certain powers to examine and approve
technology import projects has reduced the levels of administration
involved, but even at local levels many different organisations are
still involved. This devolution has been necessary as the accent on
technology import projects has moved away from a small number of
large ones to a large number of much smaller ones, which the central
agencies could not hope to administer. Under the new system, no
central approval is required for local projects paid for by locally
retained foreign exchange. Ministerial or joint ministry/local
projects still require central approval, as central funds are
involved. While the increasing degree of decentralisation has

alleviated some of the major problems associated with both the old fully centralised system and the partly decentralised system of the mid-1980s, it has brought with it new problems. The first is the strengthening of local "independent kingdoms" all acting in their own particular interests, and the continuing lack of real autonomy at the level of the enterprise. While enterprises can now retain part of their foreign exchange earnings, these are often held by their local governmen,t which can spend them on projects other than those defined by the enterprise owning the foreign exchange.

As the volume of transactions increased through the 1980s, the need was increasingly felt for a formal regulatory regime which defined the conditions under which transfers could take place. Rules introduced in 1985 laid down the valid scope and context of transfer contracts and codified the procedures to be followed in examining and approving contracts. It is unclear just how far this regulatory apparatus affected the composition and efficiency of technology flows or whether it was adhered to closely. The fact that a new set of rules was introduced in 1988 suggests that there were problems in implementing the previous regulations. The rights of examining authorities to intervene to change the content of contracts has been strengthened, which suggests a move towards recentralising control over the technology transfer process.

One aspect of the technology import process that has been the subject of increasing debate has been the manner in which feasibility studies have been carried out. It is claimed that project selection and evaluation is often done on the basis of poor feasibility studies, which leads to many problems at the implementation and assimilation stages. These studies are sometimes biased or perfunctory and are often carried out by unqualified personnel, as little attempt has been made to include relevant S&T personnel or R&D units in the process. This highlights the still largely unsolved problem of encouraging closer linkages between R&D and production units discussed earlier in relation to domestic technology transfer. There is little doubt that the low level of participation of the R&D sector in the initial decision-making has frequently the acquisition of technology inappropriate to needs and requirements and often difficult to master and assimilate.

While there is extensive criticism in the Chinese press of the still often passive role of the end-user, the enterprise which actually receives the imported technology, recent Western studies do show that the enterprise is increasingly playing a much more important role in the technology acquisition process than previously (30). These studies show that the enterprise usually initiates the process by defining the initial problems which lead to the search for new technology. This search process usually involves outside agencies such as administrators in local bureaucracies, managers in other enterprises, and patrons at municipal or provincial levels. Decisions on specific options are increasingly made by the enterprise itself, but this is done in the context of discussions held with upstream suppliers and downstream customers. In addition, a range of local agencies have to be consulted on the local ramifications of the acquired technology in such diverse areas as pollution, changed work-force requirements, taxation matters, etc. The management team of enterprises wishing to import technology therefore has to negotiated its way through a complex maze of external interests and sometimes satisfy competing demands in the process of negotiating the contract terms with the foreign supplier. This process increasingly requires managers with an entrepreneurial flair, a characteristic not normally associated with the mainstream of Chinese management.

Although it is difficult to isolate which factors are responsible for successful transfer of technology, the size of an enterprise clearly plays a role. Small enterprises with their greater decision-making autonomy often find it easier to manoeuvre their way through the bureaucratic maze and negotiate with interested parties, as the threats associated with change are usually less than those of big import projects undertaken by large enterprises, which are often subject to "blocking coalitions" imposed by threatened external agencies. The complex interactions outlined above help to explain why negotiations with foreign suppliers can take so long and how

non-technological or economic considerations can sometimes
decisively shape decision-making.

Technology transfer through foreign direct investment (FDI)

A major raison d'être for the encouragement of various forms of FDI
was the conviction that it would be an effective channel for gaining
access to advanced technology and modern management skills, and that
transfer and assimilation would be effected more efficiently and
rapidly than through other forms of transfer, as the technology
supplier, through his equity participation, has a vested interest in
ensuring success. Special measures have been included in the rules
and regulations issued since 1979 to promote FDI in advanced
technology areas. Up to 1984, however, both the overall inflow of
FDI and the economic sectors into which this investment went, were
considered disappointing. Since then and up to mid-1989 FDI has
increased rapidly, amounting to over $11 billion by the end of 1988,
and much more has gone into the industrial sector rather than into
services. One constant complaint from the Chinese authorities,
however, has been the small amount of FDI going into advanced
technology sectors. Lack of information precludes any accurate
assessment of the role FDI has played as a source of technology
transfer, though a comparison of the value of FDI with the value of
technology imports in various industrial sectors shows that the
latter far outweighed the former, indicating that FDI has played a
very modest role as a channel for the acquisition of technology.

Problems encountered in the transfer process

Very few Chinese studies are available on the economic impact of the
most recent flows of foreign technology. There is, however, a
general consensus in the specialised media that a wide range of
problems have plagued the whole process since the policy shifts of
the late 1970s, and that many of these continue to manifest

themselves today. Some data is available which indicates that many technology import projects have achieved less than perfect returns. For example, a survey in 1986 of 630 enterprises which had imported technology in the previous three years, concluded that 32 per cent had failed to commission the technology/equipment imported, only 4 per cent achieved the designed capacity while 40 per cent reached reasonable capacity levels. As far as the supply of domestic inputs is concerned, 40 per cent relied entirely on imported components while of the remaining 60 per cent, only half achieved 80-100 per cent domestic input supply (31). These figures do suggest the existence of major inefficiencies in the transfer process. The consensus is that while overall policy guidelines have been correct they have either been ignored, or implementation has deviated markedly from policy, and the mechanisms to monitor policy compliance have often been non-existent. The major shortcomings include an over-concentration on importing equipment and final assembly lines rather than disembodied technology; unplanned and unco-ordinated import; a certain amount of obsolete equipment has been purchased; import of technology inappropriate to domestic needs and conditions has occurred; the utilisation rate of much equipment is very low; there has been widespread duplication of imports; many projects have emphasized short-term economic benefits; the assimilation of imported technology has been extremely poor and slow.

In general terms, while technology import work is nominally planned at either central or local levels and carried out according to specific procedures, in practice it is claimed by some that the system is often chaotic and planless. The problem is seen as partly systemic, with rivalries and friction between functional and regional organisations in charge of co-ordinating import work. A further problem lies in the great difficulties met in co-ordinating technology import work with regional, sectoral and enterprise technology development programmes. The policy shift away from importing complete plants, assembly lines and equipment towards acquiring technology through licensing, know-how or combinations of equipment/technology has not been very successfully implemented. While the forms of technology transfer adopted are often a function of the characteristics of the particular sector in question, the volume of "software" imported in recent years has been very low, around 15 per cent in value terms. The economic pressures which operate on enterprises are partly responsible for this. Equipment imports can be quickly translated into physical production gains, whereas technology imports may realise much fewer results in the short term and may cause disruption to established production procedures. Import of technology which achieves improvements in quality, cost reductions or in products with new or improved functions, may not necessarily translate into direct economic benefits for the enterprise concerned for a number of reasons. The introduction of short-term management contract systems in many enterprises has also put great pressure on factory directors to emphasize short-term results, which works against the implementation of complex long-term technology import projects. Finally, many enterprises do not have the technical in-house expertise to implement "software" projects.

The major problem of duplication of imports is perhaps the most visible manifestation of policy failure in recent years. The phenomenon of repeated imports of the same equipment can be found in many sectors and product areas, but has been most obvious in the consumer electronic and electrical sub-sectors. The decision to meet consumer demand as rapidly as possible was achieved in several ways. The first was the large-scale import of a range of finished goods from automobiles and micro-computers to TVs, washing machines, etc. Simultaneously, domestic capacity was built up. For many products this was done by importing final assembly lines and semi or complete knock-down kits. The advantage of this strategy was that advanced products can be produced rapidly. The major disadvantage is the high costs of component and parts imports and the fact that assembly operations usually involve very little technology transfer per se, necessitating further imports of technology to manufacture component and parts and concerted localisation efforts using domestic technology and engineering resources. While there is nothing inherently wrong with pursuing such a "final product import substitution" strategy, and while policy may explicitly sanction duplicate imports of assembly lines, this has to be closely coordinated with complementary measures to upgrade domestic technological capabilities and increase local content in ways that are economically sensible and feasible.

The evidence from China is that the development of a number of product lines has been implemented extremely inefficiently, and little technological capability has been acquired. The most notorious case is that of the colour TV sector. The original strategy was to import a small number of large-scale assembly lines and factories to produce key components. This strategy quickly fell apart as the decentralisation of powers resulted in the import of over 70 assembly lines. Within a few years, China had a designed overall annual assembly capacity of over 16 million colour TVs. These lines were imported without the knowledge of the Ministry of Electronics, which was theoretically in charge of the sector's development. Few of these assembly plants could produce over 200,000 sets per year, implying no economies of scale. In addition, the wide range of foreign suppliers has posed almost insuperable problems of component standardisation and interchangeability, and few economies of scale can be developed in domestic component production. The annual import bill for parts and components is much greater than the original foreign exchange costs of the assembly lines. Because of an acute lack of foreign exchange, only part of the actual production capacity can be utilised, leading to poor returns on the original investment. Although production growth has been extremely rapid, rising from a few thousand in 1979 to 6.7 million in 1987 and over 10 million in 1988, large numbers of colour TVs have continued to be imported, despite central rules which have sought to reduce and then ban such imports. Central controls on domestic production through strict quotas on the import of key components such as colour tubes have been disregarded on a massive scale, as in 1988 over four million were illegally imported, and the central authorities appear incapable of attempting any rationalisation of the sector.

The reason for such flouting of central planning regulations by virtually every province and large city has been the perceived need to meet local demand and also the fact that inefficiant production can still be profitable, as the retail price of colour TVs is very high and the profits are largely appropriated by local governments. The response of the central planners has been a somewhat belated attempt to import complete plants for key components or negotiate joint ventures, although even at this stage they do not appear to be in full control of this process. The aim is 100 per cent local content as rapidly as possible, and great efforts have been made to start exports to recoup some of the foreign exchange expenditure for components. Despite claims of having achieved 80-85 per cent local content, it has recently been conceded that currently, this is in fact only 40-50 per cent, and that although colour TV manufacture has been listed as one of the twelve national assimilation projects, and hence given priority in terms of resource allocation, progress has been relatively slow. The pressures and incentives to increase local content in this sector at least have been low due to the high demand and the ease with which central contols can be subverted.

The next major consumer electronics product that China is to develop is the video cassette recorder sector. Again, while unified planning is being pursued to avoid the chaos of TV development, there is clear evidence that the same mistakes are being made. For example, despite strict state controls on both the import of final products, manufacturing technology and equipment, none of these have been observed. Trial production of VCRs started in the early 1980s, but by 1988 output had only reached around 40,000, mostly from CKD kits. Factories outside those few designated to do so are starting to develop production facilities. The sector's development is also being influenced by the breakdown of import controls on finished products. For example, in 1987 the imposition of an import limit of 10,000 was completely disregarded, as in practice over 300,000 were imported, with estimates that actual imports in 1988 would reach 1 million (32). A recent meeting to review the sector's development concluded that no overall plan in fact existed and current policy was in no way stimulating domestic production.

The above problems are symptomatic of those facing many other sectors, where the planning and administration of the technology import process has broken down and industrial development policy has been supplanted by short-term final product strategy. This has taken place in an environment in which each region tries to maximise its production capacity in a range of high-profit, high-demand products, with the whole process taking place in a "soft-budget" climate with little if any effective competition between producers. In this environment, entry is very easy but exit is very unusual, as inefficient producers are heavily subsidised and demand has been so great that consumers will buy second-rate products rather than go without. These factors make any moves to rationalise production extremely difficult to implement. When all industrial sectors are examined, technology imports in the past ten years have undoubtably helped to improve production capacity and, to a certain extent, technological capabilities. Chinese policy-makers have, however, on the one hand been too optimistic about their capabilities to effect rapid technical change through the import of physical plants and equipment and, on the other hand, have been unrealistic concerning the ease with which China can gain access to more "pure" forms of technology and implement it. Technology import policy has therefore tended to be formulated in isolation from overall industrial development strategy and has not taken into consideration the behaviour and interests of the actors involved in implementing policy. There is also the feeling that the overriding implicit policy objective is towards maximum technological self-reliance, despite the rhetoric of increasing global economic and technological interdependence. In an environment where the old administrative instruments of a command economy are weakened, but where new economic instruments are either weak or can be manipulated to serve narrowly defined interests, the effective management of technological development, of which acquiring foreign technology is just one component, is one of the more intractable problems facing decision-makers at all levels.

The environment for assimilating imported technology

A key determinant of the impact of the recent large flows of technology into China on the economy and on technological capabilities is the speed and efficiency with which this has been assimilated and diffused. A major problem in analysing the process of assimilation in China has been the low priority accorded to such activities until very recently, despite the lip service paid in policy statements. This has resulted in a lot of discussion but very few concrete actions, and hence in little specific information on the problems met while trying to promote assimilation activities.

The experience of other developing countries, especially the successful economic development of the newly-industrialising economies, has shown that one of the critical factors has been the management of technological development at both corporate and national levels in ways which result in efficient industrial production and investment. The key is seen to be the development of different types of domestic capabilities through the combination of different foreign and domestic technological elements in ways which ensure that a firm or country progresses efficiently along their optimum technological paths. Assimilation, i.e. the efficient and rapid mastery, adaptation and, ultimately, improvement of technology acquired from abroad through the conscious application of domestic financial, human and S&T resources, is clearly a crucial factor in this combination of foreign and domestic technological elements. Given the range of technologies acquired by a modern economy and the range of mechanisms through which flows occur, assimilation is an issue difficult to conceptualise neatly, and even more difficult to quantify.

Chinese studies on the dynamics and stages of the assimilation process came to somewhat different conclusions on the details but all essentially follow a similar, linear form, that is, import/use -- modify/improve -- independent innovation. The role of rapidly increasing local content is now a prominent feature of Chinese assimilation activities, and in some cases is seen as synonymous with assimilation. This emphasis is partly a response to the assembly strategy adopted in many sectors and the problems it has caused as outlined above. It also reflects the attitude implicit in technology transfer activities whereby assimilation is seen as a way of achieving independent innovation capabilities obliviating the need for future imports of technology in a particular area.

Assimilation is also identified with reverse engineering, which has been a major mechanism for producing new products in the past, and continues to be so in such provinces as Guangdong, which claims to have developed one-third of its new machinery in 1981-85 through imitating imported sample products (33). Although the advantages of reverse engineering are obvious, as the complexity of a product increases, so do the skills needed to be successful, as well as the costs of development and production engineering. These limiting conditions are often absent from discussions of assimilation, which is often viewed as a low-cost method of appropriating technology, whether it is embodied in imported equipment or disembodied. Such an approach does not take into account the growing evidence that most technology is specific, complex, often tacit and cumulative in its development. Part of the problem met by Chinese enterprises in assimilating imported technology is related to the fact that appropriate technological capabilities for assimilation activities are not always accumulated within the Chinese production enterprise. Because of the specificity of much technology, this expertise is not necessarily found in the formal R&D sector either. Even if the problems of co-ordination between the research and production sectors as discussed above are resolved, this may not be sufficient to mobilise the resources needed for rapid and efficient assimilation.

The current status of assimilation

Surveys of the performance of large turnkey projects imported in the 1970s and early 1980s indicate that in many cases assimilation has proceeded very slowly and often has only reached the ability to operate the plants, although in a small number of cases some investment capabilities have been developed, for example in the area of urea production. Even where this has occurred, however, the process has not been rapid and, given the much higher manning common in China, the efficiency of these plants is in question. Further evidence of slow rates of assimilation comes from a number of surveys. One, which covered 2,300 technology import contracts, found an assimilation rate of 9 per cent. Regional rates appear to vary widely, ranging from 24 per cent in Shenyang, 20 per cent in Shanghai and only 10 per cent in Beijing (34). The effective rate may be much lower, as none of the above surveys defined the extent or type of assimilation achieved.

The actual performance of technology import projects is of course a function of many variables. These include poor planning, co-ordination and management, as well as the failure to understand and master the technology acquired. A few studies on the major causes of poor performance have, however, isolated some important factors hindering assimilation. The problem of "meshing", i.e. integrating imported technology/equipment with local equipment and conditions, has been a major one, as has the lack of adequate funding. Organisational and institutional constraints, resulting in difficulties in mobilising and co-ordinating the necessary resources to apply them to assimilation problems, are also seen as an important factor. Others included lack of financial incentives and the lack of expertise at enterprise level. Successful assimilation is often a function of the backing given by local economic and S&T commissions in solving the organisational, funding and manpower problems commonly met. Other factors include the initiation of special training programmes, the inclusion of assimilation projects in local plans, and choosing projects whose products are in strong demand.

At the enterprise level, the few case studies carried out point to the following factors playing an important role. In some product areas, competitive pressures on the domestic market are forcing managers to pay more attention to the efficient utilisation and operation of imported technology than was previously the case. In some cases the greater preoccupation with incremental rather than generational technical change through technology imports is also making the task of assimilation easier. The establishment of special management task-forces to oversee the acquisition and implementation of technology import projects is considered a key to success in many projects, as are close personal links between enterprise managers and the many outside organisations involved in each transfer process. It is clear that the type of technology import strategy adopted in particular sectors has had a great impact on subsequent assimilation activities at enterprise and sectoral level. For example, starting from final assembly and working backwards presents great difficulties in assimilation and localisation work when many different brands of the same product are produced. Duplicated imports which lead to a rapid assembly capacity also greatly reduce the amount of foreign funds available for assimilation, as all parties involved have a vested interest in importing the components and parts required to keep their assembly operations going.

Administrative issues

The low priority accorded to assimilation is reflected by the fact that national regulations covering its scope, financial incentives, funding, etc. were only published in 1986. The lack of a coherent administrative framework within which assimilation
could be planned and implemented is a major cause of the slow progress made to date, as the process is often time-consuming, complex and costly, cutting across departmental, sectoral and regional boundaries. Such conditions make strong administrative leadership a prerequisite. While some regions like Shanghai set up special groups in the early 1980s, others like Beijing did not take any concerted action until only very recently. Innovations like special measures to promote assimilation and the establishment of offices to manage different projects have not in themselves been sufficient, however, to guarantee success, as attested by the case of Shanghai.

Assimilation also raises a number of administrative issues over which consensus may be difficult to reach, a situation conducive in China to taking no action at all. These include the issue of the quality and technical standards to be aimed at in increasing local content. The insistence of Volkswagen, for example, on keeping these on a par with its domestic standards has caused many problems for its potential Chinese components suppliers and slowed down the planned rate of localisation. A second issue is whether to rely on local resources or develop collaboration with "outside" units. Some evidence points to great reluctance to take the latter course of action despite the lack of suitable local facilities and expertise, for fear of a technology "leakage" and subsequent loss of competitive advantage (35). Other issues include the question of developing a division of labour and specialisation in assimilation and localisation work. Again, departmental and regional rivalries can result in the duplication of assimilation activity. Finally, there is the issue of the timing and extent of assimilation appropriate for individual projects. While the aim for maximum assimilation and localisation may be both technically feasible and economically rational for a few products, the speed at which this is to be done is dependent on a number of factors, in some cases externally located. Criticism in the Chinese press of a uniform approach to this issue indicates that decision-making in this area is often based on just a few general criteria rather than on an analysis of the specific conditions of each case.

Mobilising resources for assimilation

The difficulties of gaining access to funding is a direct result of the low priority given to assimilation until recently. While it is well known that in other countries, the costs of assimilation are often much greater than those of imports, this has not pursuaded Chinese decision-makers to allocate more funds to assimilation. Data on this issue is very scarce but some figures point to gross underfunding. For example in Shanghai, one of the cities most successful in assimilation, the ratio of technology import/assimilation funds in recent years has been 40:1, a highly skewed ratio. In 1987, this ratio for China's large-scale industrial sector was approximately 20:1 (36). The ratio is even

greater for technical renovation/assimilation funds. There are
several reasons for this low level. Assimilation often involves the
further import of technology and thus access to foreign exchange,
which is difficult to obtain. Assimilation work by enterprises may
not necessarily bring significant economic benefits, which makes
them wary of such work unless funding is guaranteed and the work
does not affect current production. Solutions proposed, but not
apparently as yet implemented, include designating a certain
proportion of technology import funds to be used for assimilation
and approving import projects only if plans for assimilation are
presented with the project document and the funds are available.

A number of financial incentives are available, such as low-interest
loans, tax breaks, subsidies and allowing costs to be passed on to
the end-user. However, these do not appear to be powerful
instruments to promote such work, which has uncertain rewards. The
current price system is a further constraint on assimilation and
localisation activities. Even when successful, sales of such
products may be low because of their high price compared to similar
imported components and products. The costs and benefits of infant
industry protection and import substitution activities are dependent
on a number of factors, but both can stimulate assimilation if
formulated and applied sensitively. China is currently pursuing
both strategies, the second much more successfully than the first.
While this may stimulate more assimilation work, the efficiency with
which this is carried out is another matter, as incentive and
penalties to move rapidly down the learning curve are often rather
weak.

The human and material resources required for assimilation may be
available within the enterprise or may involved large-scale cooperation with a range of domestic supplier, end-user and R&D units,
depending on the complexity of the technology in question and the
degree of assimilation aimed for. In the Chinese context, this
raises immediate problems of how best to organise effective linkages
between the units involved. Some Chinese commentators argue,
however, that the poor record of assimilation, including the problem
of linkages, can often be traced to decision-making process at the

technology import stage. The lack of co-ordination between the major agencies in charge of import planning and implementation, industrial technology development and R&D leads to a marginalisation of the latter in decisions on technology choices. There is also little evidence that such decisions take into account the future problems and requirements of assimilation.

In practice, most enterprises prefer to carry out all their assimilation work themselves. In a survey of 620 technology-importing enterprises, only 2 per cent had collaborated with a research institute in such work. Part of the problem is that research units may not have the required type of expertise needed for different types of assimilation work, as the knowledge is often product/process specific, resulting from production and development experience or a symbiotic relationship with suppliers/end-users. The assimilating unit may be increasingly reluctant to allow other units to participate in their work, even if they do not possess the required expertise themselves. The potential problems mount as more actors are involved in an assimilation project, as their objectives and interests may diverge significantly and be difficult to reconcile, especially if there is no clear market for the product (37). The role of administrative intervention can therefore be crucial in overcoming such problems. One organisational innovation that has been introduced to alleviate linkage problems is the so-called "dragon" model, which is a multi-disciplinary project-oriented approach to problem solving.

Although a few regions such as Shanghai announced extensive local assimilation plans in the early 1980s, little is known of how well they have been implemented. Concerted action at national level only started in 1986, when ten major "dragon" projects were approved, with two more added later. In 1987, a major change of policy was signalled which shifted enterprise technological development away from importing equipment towards the assimilation of equipment already imported. This led to the formulation of a further plan involving 150 new projects, which were to receive favourable measures in a number of areas. A recent meeting to assess progress in these two programmes has concluded that mixed results have been achieved (38). Nine of the twelve key projects have achieved initial success with some import substitution capabilities developed, but lack of funds, especially foreign exchange, was identified as the greatest constraint. Progress appeared to be much less promising in the 150 smaller projects, as no reports were submitted for two-fifths, indicating that little or no work had been done. The evident problems encountered by these State-supported projects are probably magnified at the local level, where projects have less access to resources.

Prospects for assimilation

The former neglect in assimilating the large volumes of technology imported in the 1970s and the 1980s is gradually being rectified, although the costs of this neglect have been high, and a number of factors still operate to hinder the scope, speed and efficiency of assimilation activities. Attitudinal rigidities prevent assimilation being the normal step to take after technology has been imported. Given the scarce resources involved, their mobilisation has been difficult even at State level. The economic environment, which has emphasized short-term returns, has made investment in assimilation an unattractive proposition, given the lax controls on imports. The uncertain economic returns accruing have necessitated heavy State involvement. The last year or so however has seen some policy changes which are creating a more favourable climate. Heavy dependence on finished products and components in many areas is now seen as unsustainable in the long run. Import substitution and, to a lesser extent, export promotion policies, while difficult to manage efficiently given the rather blunt tools currently available, could provide an impetus to assimilation activities, although under certain conditions they can harm overall industrial development, especially if they are operated in a hostile external economic environment which tends to reinforce a still strong technological self-sufficiency mentality. The realisation of the crucial importance of assimilation to the development of national and enterprise technological capabilities is one thing, but the creation of an appropriate administrative and economic environment in which it can take place and result in more efficient production is another. Direct State management of the assimilation process is becoming a less feasible option, apart from a small number of major projects. Although favourable economic and financial measures can contribute to influencing enterprise behaviour towards assimilation, they are, as yet, insufficient to replace heavy administrative intervention.

China's performance in managing technological development

China's strategy of rapid technological upgrading is a major pillar of industrial development policy. Its approach is based on the more effective use of domestic S&T resources coupled with greater flows of technology from abroad. Managing the assimilation of the latter plays a crucial role in combining foreign and local technological elements in ways which develop domestic capabilities in areas where they can be more efficient.

In terms of domestic S&T resource management, China has made progress in restructuring its S&T system so that the needs of industry are better catered for. This success, however, is only partial as significant problems and inefficiencies still exist in the areas of effective transfer mechanisms, funding and use of human resources. The picture is not so favourable when the direction of domestic technology flows is examined, as supply to the large-scale industrial sector, the heart of the whole system, is still limited. The major problem is the continuing weakness of demand for domestic technology by the whole industrial sector. While this is in part a function of supply problems, the main cause is the lack of incentives to adopt new technology, itself a function of the current economic environment within which many enterprises work.

The demand, however, for imported technology, especially in the form of equipment, components and assembly lines, is very strong and only limited in many cases by difficulties in access to foreign exchange. This is largely a function of domestic technology supply constraints. The superior quality, realiability and delivery times of imported equipment are also important factors, especially in those sectors whose de facto development strategy is to build up a final product manufacturing capacity as quickly as possible. Inefficiencies in technology import planning and administration, however, are reducing the benefits of the large investments made. This is especially true in the assimilation process, which has been largely ignored at all levels until very recently. This neglect has a number of ramifications. Some technology import choices are made on too narrowly defined criteria, opportunities to increase domestic technological capabilities are being forfeited, and some domestic R&D is being marginalised. The failure to emphasize the role of assimilation and to take a more integrated approach to managing technology development will make the revitalisation of China's industry a difficult objective to achieve. The dilemma is that the preconditions for adopting such a co-ordinated strategy, the continuation and deepening of economic and structural reforms, appear under the present circumstances to be further away than ever.

68

REFERENCES

1. C. Dahlman, et al., Managing Technological Development: Lessons from Newly Industrialising Countries, World Bank Staff Working Papers No. 717, Washington, D.C., IBRD, 1985.

2. G. Tidrick, Productivity Growth and Technological Change in Chinese Industry, World Bank Staff Working Papers No. 761, Washington, D.C., IBRD, 1985, p. 5.

3. For discussion of this see R. Conroy, "Technological Change and Industrial Development" in G. Young (ed.), China: Dilemmas of Modernisation, Beckenham, Croom Helm, 1985, Chapter 3.

4. This was the expression used by the famous Chinese economist Sun Yefang in an important article in Peoples Daily (in Chinese), 19th November, 1982, p. 5.

5. State Science and Technology Commission, Outline of China's S&T Policy 1988 (in Chinese), Beijing Science Publishing House, 1989, Chapter 3, p. 365. Hereafter referred to as Outline ....

6. China Daily (Business Weekly), 6th March, 1989, p. 2.

7. Modern Industry (in Chinese), No. 7, 1987, pp. 6-9 in Peoples University Printed Materials Centre (hereafter Peoples University ...), F3, Industrial Economics, No. 9, 1987, p. 138.

8. Studies in the Management of Science, No. 4, 1988 in Peoples University ..., op. cit., N1, S&T Management and Achievements, No. 10, 1988, p. 98.

9. Tidrick, op. cit., p. 54. See also his section on diffusion pp. 54-62.

10. See reference 8.

11. For a discussion on these points, see K. Pavitt, "Sectoral Patterns of Technical Change: Towards a Taxonomy and a Theory", Research Policy, Vol. 13, 1984, pp. 343-73.

12. It is interesting to note that there was no word in Chinese for technological innovation until recently, when the term jishu chuangxin has been designated. This is not surprising as innovation by its very nature implies the successful commercialisation of a new product/process in the market by entrepreneurs. The acceptance that technology is a commodity, the operation of markets and the use of technology as an instrument of competitive advantage are all new ideas in China which are still being cautiously tested.

13. Over 4,500 such enterprises had established technology development organisations by 1987, of which 30 per cent were set up in 1986 and 1987. Outline ..., op. cit., p. 366.

14. Ibid., pp. 39-41.

15. Management World (in Chinese), No. 5, 1987 in Peoples University ..., op. cit., N1, S&T Management and Achievements, No. 12, 1987, p. 78.

16. These figures are given in Outline ..., op. cit., Section 1, Chapter 2. Note that these include technology transactions in all economic sectors. The contract value for industrially-related technology accounted for 64 per cent of total contract value in 1987.

17. Outline ..., op. cit., Tables 2.13, 2.15, p. 350. Note these figures are for all S&T activities, not just industrially-related ones. Income of all industrially related R&D units came to 52 per cent of total income.

18. Outline ..., op. cit., p. 367.

19. Science and Technology Daily (in Chinese), 5th October, 1987, p. 3.

20. Outline ..., op. cit., p. 377.

21. China National Research Centre for S&T for Development, A Collection of Reports on the Reform of the S&T System (in Chinese), Beijing, 1986, p. 3.

22. The rest of this section is drawn from G. Tidrick and Chen Juyuan (eds.), China's Industrial Reform, New York, Oxford University Press, 1987, passim. Additional observations are taken from current work being done by the author on factors influencing demand for technology by industrial units.

23. For a historical review of the different phases of technology imports, see "A summary of our technology and equipment imports since 1949", in Economic Research Reference Materials (in Chinese), No. 186, 1984, pp. 9-17.

24. Chen Huiqin, "The orientation of technology import must be changed", Economic Management (in Chinese), No. 4, 1981.

25. China Economic News, No. 3, 1987, p. 1.

26. The following material on Shanghai is taken from Peoples Daily (in Chinese), 20th August, 1988, p. 3; Science and Technology Daily, 5th January, 1989, p. 1; China Economic News, 23rd January, 1989, p. 12; China Daily (Shanghai Focus), 13th March, 1989, p. 2.

27. The data for Guangdong is taken from New China News Agency -- English, 5th May, 1988 in BBC Summary of World Broadcasts, Part 3, The Far East, W0026/A/5; Liu Wenyan and Shi Lei, "An analysis of the characteristics of Guangdong's technology imports", Imports (in Chinese), No. 1, 1987, pp. 68-9 in Peoples University ..., op. cit., F3, Industrial Economics, No. 4, 1987, pp. 79-82.

28. China Economic News, 15th September, 1986, p. 14.

29. For the measures introduced in mid-1989, see China Economic News, 12th June, 1989, pp. 2-3.

30. Roy Grow has been dominant in such studies. See for example, R.F. Grow, "Acquiring Foreign Technology: What Makes the Transfer Process Work" in D. Simon, M. Goldman (eds.), Science and Technology in Post-Mao China, Cambridge, Mass., Harvard University Press, 1989, Chapter 13.

31. Bai Yiyan, Some Policy Issues of Technology Transfer to China, paper presented at the workshop "Technology Transfer to China: An Assessment and Revaluation", Tufts University, US, March 1989.

32. Peoples Daily, 21st October, 1988, p. 2.

33. Impact and Consultancy (in Chinese), No. 1, 1988, p. 9.

34. Research Management (in Chinese), No. 2, 1987, p. 14; Peoples Daily, 20th August, 1988, p. 3; New China News Agency, 29th February, 1988, in BBC Summary of World Broadcasts, Part 3, The Far East, FE/OO89/B2/5.

35. Cf. Industrial Economics and Management (in Chinese), No. 5, 1987, pp. 40-43 in Peoples University ..., op. cit., F3 Industrial Economics, No. 7, 1987, pp. 95-98. See also Economic Information (in Chinese), 27th, 28th and 29th July, 1987, p. 1.

36. For the first figure see Peoples Daily, 20th August, 1988, p. 3. For the second see Outline ..., op. cit., p. 377.

37. For such problems see Research Management, No. 2, 1989, pp. 14-18.

38. For a report of the meeting see Science and Technology Daily, 12th August, 1989, p. 2.

FROM TECHNOLOGY TRANSFER TO TECHNOLOGY MANAGEMENT

Pierre Ventadour
Lecturer, Le Havre University - International Affairs Faculty
CODASIE - Tour Générale - 5 Place de la Pyramide
92088 PARIS LA DEFENSE

Whenever the Chinese refer to the situation in their country in the field of technology - in books of history and political economy, in newspapers or on the radio - they systematically and quite officially refer to it as being luo hou, which means backward or lagging behind. With this expression, we must allow for traditional Chinese politeness, which involves the use of ecxessively derogatory words when one speaks about oneself, and the use of excessively positive and favorable words when talking about someone present. Yet there is something else: a general feeling, deeply rooted in people's minds, that foreign knowledge is superior to their own in the fields of science and technology.

The origins of this attitude must be traced back in China's history and culture, particularly in the developments of the last 150 years. Ever since its beginnings, Chinese civilization has had the Confucian philosophy as its single and permanent mainstay. This way of thinking has generated a concrete way of looking at the world; the Chinese consider the universe to be a large mechanical and biological whole, which rotates and changes. This has nothing in common with the Greek and Judaeo-Christian outlook, which gives priority to abstraction and rationalism, and which gave rise to the mathematical model concept.

Confucian philosophy has permitted the continuity of Chinese civilization, which has suffered a break since prehistorical times. The Mongol episode, which could have resulted in the elimination of China as a nation, and which caused the death of about half of China's total population, left the Chinese people with a deeply rooted unconscious fear of anything foreign. Under the Ming dynasty, China cut itself off completely, and in the 18th century developed a civilization which

Europe-Asia-Pacific Studies in Economy and Technology
Leuenberger (Ed.) From Technology Transfer
to Technology Management in China
© Springer-Verlag Berlin Heidelberg 1990

turned it into a haven of peace and prosperity. In the early 19th century, prior to the assaults of the western powers during the wars to force the opening of certain ports, China experienced a domestic economic disequilibrium, due to a growth in the population rate that was in excess of agricultural output, so that China was no longer capable of feeding its population. This domestic weakness, as well as the shock wave of wars, caused China to retreat on the military, economic and cultural fronts before the west, which, employing modern sciences based on the development of methematics, physics and chemistry, had built both the weapons and tools that gave it military and economic power.

After the Manchu dynasty was overturned and a republic proclaimed in 1911, the Chinese intelligentsia concluded that the one and only solution for China to become a modern and powerful nation was to adopt foreign ways of thinking in the fields of science and technology. From 1915 onward, a group of progressive intellectuals, led by Hu Shi, as well as by Chen Du Wiu and Li Da Zhao, who were later to be among the leaders of the Chinese communist party, came to the conclusion that China must completely and finally turn away from the Confucian way of thought, and adopt western thinking patterns.

While the success of the Chinese revolution in 1949 restored China's dignity and sovereignty, it resulted in China closing up and cutting itself off from the outside world for 30 years. Even today, an ambivalence remains in the minds of many Chinese people as to the appreciation of foreign ways of thinking and of foreign products. Such an appreciation is rarely unbiased, and the Chinese often feel either an excessive attraction or an excessive reluctance with regard to things foreign. It may be argued that Chinese judgement in the field of technology is an illustration of such excessive and inappropriate attitudes.

Knowledge of the situation as it exists on the spot, and experience of technical cooperation in China, are ample evidence that the technological gap is not the major obstacle to China's economic development. First of all, this gap is neither as wide nor as real as the Chinese say or believe. Also, countries such as Taiwan, Korea, Hong Kong and

Singapore, whose economic prosperity and adaptability as quoted as examples, do not have high technology comparable to that developed by western countries; indeed, they are far from being abreast with China in some fields. In fact, China has a degree of know-how, and in any case a potential which is lacking in many of the countries which approach the international markets successfully. Let us analyze the situation in detail, and try to explain what the obstacles are that China has to overcome.

* * *

Western scientists and researchers from major research agencies or major industrial corporations who have visited Chinese laboratories, say in most cases that they were impressed by the quality and often the quantity of the testing equipment, as well as by the quality and the number of computer facilities available in both the basic and applied research laboratories that they visited. This wealth of state-of-the-art equipment compares rather too favorably with the cuts often imposed on European laboratories; indeed, there is probably a happy medium between these two situations. The amounts expended by Chinese laboratories were in fact often excessive.

Visits to institutions in Beijing, Hefei in Anhui province, Shanghai and Guangzhou, too, showed that a number of research laboratories had worldwide recognition for the quality of their research and for the number and quality of their publications. Some of these institutions even ranked among the world's best three laboratories in their specialized areas. Shenyang, the capital city of Liaoning province in Manchuria, also deserves special mention: a highly reputed research center on robotics has been established there. Finally, Quinghua University impresses visitors with the quality and standards of both its teaching and the research done in its laboratories.

Most of the Chinese students or researchers who come to study or work in western laboratories win the respect, esteem and even the admiration of their western counterparts, due to their intellectual abilities, the quality of their studies and training, as well as by their eagerness, their application at work, and their efficiency. We might be tempted to believe that this is because they had all these qualities in the first place, that they were handpicked for two reasons: on the one hand, to ensure that they would benefit as much as possible from the training and experience they would have abroad, and on the other hand to boost China's image abroad by sending people of outstanding quality, capable of impressing the teams they go to work with.

This assumption has not been confirmed by facts. When Chinese researchers and technicians travel in groups, the quality of the team is homogeneous, and the standards are very high. Also, when one travels through China and visits universities, laboratories or industrial concerns, one is impressed by the general standard that the people show, irrespective of whether they have travelled, studied or stayed abroad or not.

We must therefore conclude that the standard of Chinese researchers and technicians is quite comparable with that of their western counterparts, that the equipment available in laboratories is for the most part state-of-the-art, and also that these laboratories often have more facilities than research centers in other countries. In many fields, the quality of the research and achievements by Chinese laboratories and industries at the spearhead of technology is very high; in some cases, it is even the best in the world. China, despite the setback of its highly troubled history over the last 150 years, is now capable of meeting its requirements in the high-tech sector quite independently.

China has constructed nuclear weapons, an H-bomb in particular. Scientists such as Mr Qian San Qiang, who was trained in France, where he lived from 1937 to 1948 as a student of Joliot Curie, and others trained in the US, brought together teams, made the equipment, and measured and tested the devices they required. In a difficult domestic political and economic environment, without any outside help and virtually cut off from the world, they successfully constructed the atomic weapons China had decided it wanted to possess.

With a rocket called "Long Walk", China now has a launcher that has successfully placed teletransmission geostationary satellites in orbit. The performance of this launcher is sufficiently efficient to encourage China to offer its services on the international market. Even if the performance of the "Long Walk" rocket does not match Ariane's, particularly with regard to the payload, and if it is not so sophisticated as the European launcher, we can only be impressed by the Chinese performance, compared with the achievement of a manufacturer which combines the competence of several European countries which are considered to be spearheading world technology.

In the field of aeronautics, China is in the process of building a small all-Chinese aircraft, whose performance compares favorably with other similar planes. In biology and biotechnology, as well as in a number of life sciences, the quality of the research and findings of Chinese laboratories places China among the world leaders. In basic and applied research and in the high-tech industries, China has shown that it has the necessary specialist personnel, that the laboratories have high-quality testing, measuring and control equipment in adequate quantities, and that Chinese achievements in many fields compare with the best foreign achievements. It is therefore possible to say that in these fields, China is neither at a disadvanage nor lagging behind.

Chinese specialists and technicians, whether they are on industrial or training visits abroad or playing the role of host at home, also show that they are conversant with the latest industrial techniques used throughout the world, that they understand the pros and cons of these techniques, and that they are interested in the most up-to-date and most efficient equipment available in their particular fields. Apart from the obvious advantages of this, there are also snags. The following example may serve to illustrate this. A Chinese engineer was required to build a liquefaction plant and the power supply facilities to go with it. It turned out that he had designed this project by carefully selecting the best individual parts available in the world, not taking into account the compatibility and connection problems, say, between motors and pumps or between turbines and generators. The resulting heterogeneous system did not perform as well and efficiently as a homogeneous one would have done, and maintenance costs were higher, too - as were the acquisition and installation costs, which exceeded those for a homogeneous system quite substantially. This is a significant example showing one of the major difficulties facing China in the implementation of its economic reforms and of its policy of opening up to the outside world.

Buying the best individual components without always paying attention to the best possible end result, is not only a consequence of the fact that Chinese engineers are aware of the latest equipment available on the world market, but may also be due to the elitist nature of Chinese education. Then again, foreign suppliers approaching the Chinese market do not always integrate these factors in their approach - either because they are unaware of the characteristics of this specific market, or because they are out to achieve the highest possible sales figures by selling the most expensive equipment, regardless of whether this is suited to the Chinese users' requirements or not.

It may also be the case that the Chinese themselves put excessive emphasis on the purchase price, and not enough on operating and maintenance costs, on the yield and the life of the equipment. The Chinese tend to choose the least expensive offer. Some of the foreign suppliers, rather than insisting on the technical advantages of the equipment they offer, will propose low-cost equipment to win the contract, although they are fully aware that this equipment does not meet the requirements. Once the order has been placed, they then gradually get their Chinese buyers to accept a number of alterations, for which they are charged. The Chinese, then, show an ambivalence with regard to technology: they give preference, either to technical performance regardless of price, or to price at the expense of technical performance.

The cement industry, too, provides an interesting example of the weak and strong points often characteristic of the Chinese economy. Chinese cement factories are usually much smaller than European ones; for the most part, they use wet-process technology. Compared with the dry process now in general use in most western plants, this technology has the inconvenience of consuming more energy per ton produced.

Nonetheless, it enables the processing of cement of a fair average quality without undue difficulties. Consequently, the Chinese cement factories are perfectly capable of meeting the domestic market requirements for the most commonly used cement qualities, even if more energy is consumed in the process. Today, China is the world's largest cement producer and consumer, with a production and consumption estimated at about 190 million tons (this figure must be taken with reservations, as such estimates have a fairly wide margin of error).

The construction of hydroelectric dams or nuclear power stations, however, requires high-quality cement with specific characteristics: while setting, such cements must release a minimum amount of heat so as to avoid cracks, which would of course reduce the strength of arch-gravity dams and the structure of nuclear plants. Although Chinese cement factories have the technology necessary to manufacture this type of cement, they are not generally capable of producing large quantities of special cement of a consistent quality. In consequence, the cement currently being used to build the Daya Bay nuclear power plant is imported. Even so, China has already built large dams with Chinese cement: the dam on the Wu Jiane River, for example, which is a 180m high arch-gravity dam built exclusively with Chinese materials, labor and equipment; the alternators were manufactured in the Dong Feng plant.

This illustrates a major point relevant to the objective assessment of the essential requirements of the Chinese economy in terms of its economic policy and its policy of opening up. What China lacks most desperately at the moment is not technology or modern equipment; as mentioned above, China is in fact quite capable of carrying out major projects in areas that are at the forefront of modern technology - atomic energy, aerospace, aeronautics - and of manufacturing the fundamental industrial equipment needed by the Chinese economy. What China lacks is the capability of mass-producing such equipment, of maintaining quality according to international standards, and of achieving cost prices which make the products competitive on the international markets, as well as the ability to use their equipment in a productive and cost-effective manner and, in particular, to maintain in good working order over a long period of time.

Visitors to the factories that constitute China's basic industrial fabric - the ordinary factories which are the most numerous - will note a number of features that these have in common. For one thing, the equipment in use is usually old and does not enable the manufacture of goods that are up to today's international standards. Only a few years ago, however, production machinery dating back to the First World War could also be found in plants of major manufacturers famous worldwide for the quality of their products and the reliability of the production lines. Customers never complained about the quality of the goods produced by such iron veterans, whose appearance and identification plate gave away their age, but which looked almost new. Such machines had been carefully serviced for half a century, worn parts had been changed in due time, they were properly greased, oiled, and used with care. By contrast, Chinese production machines are in bad condition most of the time, and badly serviced. Individual parts of the machines fall prey to corrosion. The machines are not greased well, or with sufficient care, or at all. Alternatively, they are greased too much. In January 1989, a fire broke out in an industrial complex in Guizhou province - due to excess grease put on one machine by a team of oilers.

Thus, if the Chinese leaders claim that the old age of the equipment in use in local industries is one of the main problems China has to overcome, then this is definitely true. It must be added, however, that there are western companies which manufacture quality goods, which are then marketed under reputed trade marks on the international market, with equipment which is just as old, if not older, than that used by Chinese industries. In my opinion, the main problem is not the age of the equipment but the way in which it is used and maintained. In fact, the Chinese would be capable of solving these problems now, and with no great strain on the foreign currencies they are so much in need of; possible solutions will be indicated in the last part of this paper.

The second aspect is the usually barely rational layout of the production plants, of industrial premises in general, as well as of storage areas for raw materials, semi-finished and finished goods. In a plant in Shenyang, for instance, access to the coal store is such that if a

mere three lorries are simultaneously driving in through the yard, un-
loading, or leaving, this causes a jam that holds up everything.   If
further lorries arrive, they have to wait in a queue, which makes
matters even worse. The delivery pace is thus slowed down to a considerable extent. Whenever the situation gets completely stuck, however, it is resolved by a passing foreman or worker, who spontaneously
directs the traffic until the problem is resolved.   What is dealt
with, though, is the traffic jam, not its actual cause: in fact, the
management of this particular plant never thought of altering the
access layout to avoid bottlenecks, or of putting a man on traffic
duty to control access to the yard and prevent jams - indeed, the
question had not been given as much as a thought.

Similarly, the layout of workshops involved in the manufacturing process is hardly every designed in a rational manner. One gets the impression that it is the end result of the various phases of the
plant's extension, and that there has never been an overall survey to
try and improve general functionality and efficiency.

The layout of workshops is also responsible for a considerable waste
of energy - which is all the more detrimental as China is desperately
short of this commodity. Going from one workshop to another - often
because products in progress are handled by workmen pushing them along
on trolleys - usually involves leaving the building. In areas like
Manchuria, where temperatures frequently fall to -15°C or even -20°C,
such constant movements in and out of workshops, which are often only
closed by tarpaulin, causes a substantial waste of heat. On most industrial sites in China, each factory is forced to be idle one day
every week, to ensure that electric power requirements do not exceed
the maximum power provided, since otherwise there would be power failures that would have an adverse effect on the whole site. In part,
such restrictions are due to waste in other areas. To combat waste,
the workshop temperature is kept down to nearly 0°C. Workers are
wrapped up in warm clothes, which often restrict their movements.
Both the energy and the tension created by work in these unfavorable
conditions are wasted, and last but not least, the cold makes tuning
and adjusting operations more difficult.

The third factor usually pointed out by visitors to Chinese plants is the fairly low utilization rate of the equipment. Quite often in Chinese workshops, over half of the machines are idle. There are several reasons for this: in some cases, the machines are not in use owing to lack of spares, sometimes it is due to inadequate production planning or irrational management. Beijing taxis are a case in point: many are ready in front of hotels, with the driver at the wheel waiting for passengers; other are empty and locked up. The use value of these taxis decreases fairly quickly in the course of time, irrespective of whether they are used or not; there is a loss of invested capital, which is all the greater if the assets are not properly employed. This is an example where China has surplus equipment, as well as the personnel to make use of it, but irrational practices and inadequate management.

The fourth factor which characterizes Chinese companies is a lack of adequate quality control. This does not mean to say that Chinese products are never of good quality - indeed, some products are of excellent quality; it simply means that the concept of quality control as an essential function within the production process does not exist in the majority of Chinese plants. Of course, industrialists are trying to manufacture in better conditions, but they seldom enforce quality controls, either on random samples or on each unit produced. In this manner, they cannot evaluate the measures implemented to upgrade quality, or reject parts which are not up to quality standards: output is output, whether high grade or low grade, and the products are put on the market as they leave the production plants, good or bad.

This is made possible by the distribution and marketing channels of the Chinese market. The book trade provides an informative example. If you want to buy a book in Beijing, whose bookshops are as a rule well stocked with very interesting books and magazines, you may first of all thumb through all the pages of any sample copy. Once you have chosen a book, you go and tell an assistant, who then fills in a form specifying the title of the book, as well as its price. You then proceed to the cash desk to pay, and return to the assistant with the form duly rubberstamped by the cashier. The assistant will now hand you a copy of the book you have asked for. You may not choose the copy yourself, and you will often be presented with a book with faulty binding or printing. Any complaint would be to no avail. In such circumstances, the publishers may continue to market their products, totally irrespective of the quality of the work done by the printing press.

Another example where a lack of consistent quality is detrimental to the Chinese economy, is provided by the cement industry. As mentioned above, Chinese cement manufacturers are capable of producing high-quality cement, even special cements such as those required for the construction of hydroelectric dams or nuclear power plants. Chinese manufacturers, however, are not capable of producing such cements in large quantities of consistent quality. Thus, although it has the requisite know-how, China is compelled to buy cement abroad at the cost of important amounts of foreign currencies.

The fifth factor that strikes western businessmen who are used to working in a competitive environment where the quality of products is always appreciated in the light of the selling price, is that Chinese firms have no operational cost price. This is definitely not due to the lack of qualified accountants; on the contrary, such personnel are often very well qualified. Top-management staff involved in a Chinese-French joint venture in Liaoing province in Manchuria, for instance, realized that Chinese accountants were every bit as competent as European ones. The problem is not one of qualifications or of inadequate techniques; rather, the difficulties experienced by Chinese companies in determining the cost price of their products, are linked to the specific features of the economic system that China has developed ever since the establishment of the People's Republic in 1949. It is a system in which the economy is administered, not managed. The reasons for which China opted for this type of economic system are more complex than most westerners think; western businessmen and economists generally believe that the choice is a direct consequence of the political system introduced by the Chinese communists.

It must be emphasized, however, that even though the formulae may have changed, China has always proclaimed that it put into practice "a socialism Chinese style" (zhong go te se sue hui zhu yi). The Chinese

economy used to be planned and collectivist before. From 450 to 400
B.C. the Wei Kingdom, formed in 450 following the division of the Jin
Kingdom into three territories, became strong and rapidly developed
its wealth and power by applying, among other reforms, measures typical of a collective and planned economy. Thus the administration of
the Kingdom bought grain on the basis of a fixed price to build stocks
in surplus years, in order to sell it at the same price in years of
shortage. This price regularization technique is contradictory to an
attitude which would have left the market forces free play during the
time of shortage.

At about the same period, in the realm of Qi, close to the territory
of the Shandong peninsula, the administration of the Tians - liegemen
of Qi - marketed wood and fish at the same price for everyone throughout their own domain, irrespective of the transport costs involved.
The same administration used greater than standard measures to distri-
bute grain to peasants in times of shortage, while the measures were
smaller than standard when peasants came to return the grain they had
borrowed. During the empire, merchants were regarded with suspicion,
and often their rights were not identical with those of the rest of
the population; in particular, there were times when their children
did not have the same rights as regards education.

Ever since prehistoric times, China has been a nation of peasants.
Trade never really existed in the form it took at the beginning of
western civilization, where it has been one of the main vehicles and
chief components of society. This shows that the current situation is
not only due to the implementation of a socialist system since 1949,
but also to China's history, and to the development of the Chinese
civilization.

This is a fundamental aspect. For thousands of years, the tree of
Chinese civilization has grown from its roots, which are different
from the roots that have fed the tree, or rather the trees, of western
civilization. For this reason, any attempted graft of a twig from
western trees onto the Chinese trunk has been rejected. This is a
fact that neither those Chinese apt to be carried away by an excessive

enthusiasm for the west, nor those westerners who are prisoners to their egocentric views, must forget if want to avoid a nasty shock.

China's failure to implement a strategy of expanding its economic policy, then, is not due to actual technological problems, but to methods, organizational questions, reflexes and attitudes preventing the use of modern technologies in an efficient and profitable manner. This does not mean that today China has all the technologies it requires in all domains, although it has been seen that in some high-tech sectors, China has been capable of developing the requisite know-how by itself. Yet if China allocates substantial resources to the acquisition of technology, without carrying out the changes necessary for this technology to be used in an efficient and profitable way, such efforts and investments are to a large extent wasted, particularly with regard to the foreign currencies that the country is currently short of. What China needs most today is to learn to manage modern technologies, which are in essence foreign to Chinese philosophy, and whose use has been designed in an economic system which China has never experienced in its entire history. The real problem facing both China and the western countries wishing to develop lasting and balanced economic relations with China, is to achieve a successful graft of a western twig onto the Chinese tree.

TENTATIVE SOLUTIONS

When the Chinese talk about foreign-made tools or foreign-designed systems which they are using or have borrowed, they are always extremely careful to emphasize the specific Chinese character. This is a sign of China's determination not to abandon or lose its specific culture and civilization, but it is also a reflection of the basic otherness of its civilization, as well as of the need permanently to recall to both the Chinese themselves and to foreigners that a phenomenon of rejection still exists.

Even so, there are numerous Chinese today who believe that the solution to the problems their country is experiencing lies in the implementation of the western system in the fields of politics, philosophy, welfare, science and economics. This Chinese attitude takes its bearings from an attitude very common among westerners who often do not have an intimate knowledge of China, and whose only points of reference are those provided by their own western cultural and educational backgrounds. Both Chinese and westerners thinking along those lines make the same mistake, though for different reasons, assuming as they do that China is on the right track if it follows the path opened by the west, and that it goes off course if it leaves this path. This mistake, quite common now, used to be made in earlier periods, and evidence has shown that it cannot be a solution to China's problems.

At the turn of the century, during the period which followed the opening crisis and the humiliation of China before foreigners, part of the Chinese intelligentsia examined the reasons for western superiority in the military and economic fields, and decided that its origins lay in the western thought which had introduced physics and chemistry into the sciences, and democracy into politics. Part of this intelligentsia felt that China would only have to introduce and apply foreign models in order to achieve both "military strength and economic wealth". The first stage was the proclamation of a Republic in 1911, subsequent to the overturn of the Manchu dynasty. The idea of a Republic did not, however, take root, and the death of Yvan Shikai put an end to attempts at restoring the empire and sparked off a process of division and unrest which was quite traditional in China's history.

The second stage, which started in 1915, was characterized by the iconoclastic attitude whose exponents were Hu Shi, Li Dazhao and Chen Duxiu. The progressist fraction of the Chinese intelligentsia, consisting of lecturers, writers and scientists, completely and utterly rejected the Confucian philosophy which had underpinned the Chinese soul for almost 4,000 years, and enthusiastically turned to the scientific, democratic and individualistic ideas of the west.

From then on, the most radical elements, led by Chen Duxiu and Li Dazhao, advocated the most recent western philosophy, which was also considered the most modern at the time: Marxism. The Chinese Communist Party was founded in 1921, expanded at an extremely rapid pace, and sparked off the proletarian revolutionary process along the lines of the Soviet pattern, which in the eyes of the people had just passed a concrete test with the success of the October Revolution. This western approach, however, was strongly rejected in 1927 with the sacrifice of the Shanghai martyrs.

Soon the revolution was on its way again, this time following the specifically Chinese tracks through the farming areas. Mao Zedong and his team followed a time-honored path and led the Chinese people to victory in 1949, though relapsing into the old Chinese reaction of cutting themselves off from the rest of the world. The Chinese, as well as the westerners, had in those days forgotten that it is quite impossible to try to implant an essentially different way of thinking without causing a rejection. This is still true today; and if this fact is disregarded, the failures of the past will be repeated in the future. Such mistakes are much easier to make even now than many believe - and in many different fields, such as in the arts, the way of life in general, politics, and in the economy.

As far as the economy is concerned, China's present-day problem is to introduce into traditional attitudes and environments whatever is necessary for the Chinese to be able to make efficient and profitable use of technologies, equipment and macro-economic systems which have not been born out of the Chinese way of thinking, and which as a result do not naturally fit in with the behavior of Chinese people. Consequently, there is necessarily a gigantic need for training and education, since what is at stake for the Chinese is learning how to use, and how to incorporate into their way of life and their social and economic environment, something that has grown on an entirely different soil.

Aware of this need, the Chinese government accounted for it in its policy, which was beginning to be implemented in 1978, of reforming the economy and opening up the country to the outside world. China decided to send abroad a number of students chosen from among high-fliers. These students were to meant to complete full courses at universities and technical institutes in a number of countries including the United States, Japan, West Germany, France, Switzerland and the Scandinavian countries. They had just left school - i.e. they were about 17 to 18 years old - and completed a one-year language course to become sufficiently fluent in the languagein which they were going to study and take exams, including competitive exams.

If we take the French example of a Chinese student taking the matriculation exam at one of the major engineering colleges, the curriculum was usually as follows: a one-year French language course in a language center in Toulouse, Aix-en-Provence or Besançon; one preparatory year in advanced mathematical studies, one or two years in special mathematical studies; two or three years in an engineering college followed by two years of practical studies. The curriculum could run for a maximum of nine years. At the same time, some Chinese students do a B.Sc. course, take a Master's degree or a Ph.D., which may increase the duration of their studies by a further two to three years. Moreover, some students will spend two or three years in a third country, and will therefore be away from China for over ten years. This has numerous consequences.

First of all, these students leave China when they are in their adolescence, with no experience as adult citizens. They only know of their native country what they saw during childhood and adolescence; they know virtually nothing about Chinese society and industry. For more than ten years, they are immersed in a society to which they will gradually adapt, and their way of thinking, which had not had time to be shaped in the Chinese mold, will conform to that of the west. Their way of life will be that of the west, and their ties with friends or through marriage will put them in a situation that makes a return to China more difficult.

Some will settle in their host countries; but those who return to China face a dual problem. First, they are not familiar with the Chinese industrial fabric, and they are not aware of the specific characteristics of the Chinese economy. Impregnated by western culture and education, and molded by their experience in the west, they will tend to have an excessively critical attitude toward the environment in which they live and work; most of the time, they will be tempted to recommend or apply systems which are well suited to a western environment, but hardly adaptable to China. There will not be a proper intellectual process to harmonize the knowledge acquired abroad and the need to modernize the Chinese economy. As a consequence, the rejection phenomenon mentioned above cannot fail to occur.

The Chinese government is aware of these difficulties, and has therefore altered its general policy: fewer students are sent abroad who are at the beginning of their university careers. It may indeed be advisable for Chinese students to follow the entire curriculum in China, learning one or two foreign languages in the process: first, English, which as an international language will give them the possibility of reading a large number of books and magazines in their fields of interest; second, the language of the country where they plan to go at a later stage to add to their knowledge.

Once they have finished their studies, the students - who are now qualified engineers - will start their career in China. They will be able to assess in depth not only the weak points of the Chinese economy, but also its strong, and potentially strong, points. With this local knowledge and experience, young engineers between, say, 32 and 38, may go abroad, either to write a Ph.D. thesis on a subject related to their research work, while working in an industry, or to work in a western company for a period of six months to two years in order to become familiar with different techniques and organizations.

In such circumstances, Chinese engineers will be able to choose their subject matter, to direct useful research that takes the requirements of their country's economy into account; they no longer act as passive students but as acting executives. Moreover, such young Chinese executives will start professional and personal relations with their western counterparts, which they may later on use to keep in touch with foreign companies. Also, the intimate knowledge of the Chinese economy they have acquired on the spot will enable them to provide their western counterparts with information about China that will prove useful for purposes of comparing the general economic environment in general and individual aspects of it in particular.

A second approach enabling China to integrate western technology while taking its own specific requirements into account, would be to invite western executives and technicians into Chinese industries. To have a chance of success, these executives and technicians must have both a thorough and comprehensive experience in their fields of competence, and a sufficient understanding of Chinese reality. Both are essential. A lack of specialist knowledge involves the risk of not being taken seriously, since Chinese engineers have a comprehensive, if not universal, knowledge of their fields; a lack of field experience in working conditions poses a danger of being too dogmatic or theoretical for the Chinese, who are by nature practical and pragmatic people. Finally, westerners who are not sufficiently aware of Chinese reality, run the risk of being rejected altogether.

The basis of the Chinese way of thinking and mentality is in its essence so totally different from its western counterpart, that westerners must make substantial efforts to make understanding at all possible. After a while, western executives and technicians will have acquired concrete experiences which, combined with their initial knowledge, will enable them to make a precise assessment of Chinese requirements and constraints, and to decide in favor of the solution that can be adapted best. They will also have a real exchange of ideas and experience with Chinese managers who have been trained abroad; this, too, can only be beneficial.

What is lacking most in China is not technology itself - in most cases, the Chinese are perfectly capable of devising the technologies they need - nor is the country short of technicians that are as highly qualified as their western counterparts. The problem is the economic and social background, i.e. the interface between the new technologies and the Chinese environment, an interface which in itself is made up of organizations and processes. An analysis of the circumstances and reasons underlying the prosperity of Chinese countries such as Hong Kong, Taiwan and Singapore, or of countries with a Chinese background culture such as Japan and South Korea, will corroborate this point.

After its defeat in 1945, Japan was placed under the supervision of the United States, with General MacArthur in charge. For several years, Japanese tradition and US technological progress were allowed to blend, and this accelerated the creation of modern Japan. South

Korea, too, benefited from the advance of the United States in the fields of technology and organization. During the Korean War and the ensuing period, US experts contributed toward the creation of the modern and competitive industry in the various sectors of the Korean economy. In many ways, the major international Korean groups follow the pattern of their US counterparts. Later, Japan made its imprint on Korea, too.

Among Chinese countries, Taiwan first benefited from US aid, and then from Japan's example; the family or business ties which were established between Taiwan industrialists and US industrialists of Chinese origin, facilitated further exchanges and contributions. Hong Kong and Singapore are two of the greatest crossroads of international economic flows, and the two cultures merged smoothly at the pace of business expansion. The example of these five countries, steeped in Chinese culture, which succeeded in achieving an efficient and flexible economy by rapidly tuning in to changes in the world economy, is significant and shows that in each individual case, there has not been a mere transfer of technology but instead, two-way contributions and a harmonious integration accompanied by profound and lasting exchanges with the west.

Although China may be the source of the main components of these countries' cultures, the problems it has to solve are different in nature, not least because of the country's actual size. Nonetheless, China will have to switch from the concept of technology transfer to the concept of modern technology management. This has been understood by the Chinese leaders ever since 1979, and has found expression in the two major thrusts of their policy: the reformation of the economy and an opening-up toward the outside world. The Chinese leaders' task, however, is much more complex than that faced by the political leaders in Taiwan, Singapore, Japan and Korea.

Primarily, China has a size problem. A country that is 3,400 miles long and 3,100 miles wide, China has to overcome its huge distances; this creates a tremendous need for bridges and highways - indeed, the great distances are an obstacle to the development of exchanges within the country itself. Then, there are numerous local differences. For more than 4,000 years, China's history has been conditioned by the balance between the empire's central powers and the centrifugal forces created by specific regional features such as the large variety of dialects - according to a Chinese saying, people do not understand one another if they live more than thirty miles apart. Since the creation of the People's Republic and the teaching of Puthongua, "the common language", which is taught in primary schools throughout China, Chinese people from different provinces can now talk to, and understand, each other - in many cases, however, still only in theory.

China also faces an excessive birth rate problem. It has a population of over 1,100 million, and the growth rate - particularly in rural areas, where 80% of the overall population still live today - remains largely in excess of targets set by the Chinese government. Unless China succeeds in putting a brake on the increase of its population, it runs the risk that despite all efforts made, the annual increment of the Chinese population will consume the annual increase in goods produced, so that the Chinese standard of living would stagnate. It also runs the risk of upsetting the balance between man and nature, of damaging the environment, and eventually causing a decrease of farming output.

Moreover, China is currently undergoing a period of very rapid development, and this implies the risk of destabilizing Chinese society. In a policy aiming at the reformation of the economy and at opening up the country to the outside world, both the price and incomes structure will necessarily be subject to prfound changes which, at least temporarily, may have adverse effects on social relations and on Chinese society and the economy. The monthly salaries of Chinese top executives are currently in the 200-500 yuan bracket (there are approximately 3.7 yuan to one US dollar). Such executives have a company flat, for which they pay a monthly rent of 12 yuan. Today, under a government policy in favor of home-buying to speed up the construction of flats and, in particular, to alter the financing schemes for housing, some flats have been put on the market in Beijing, which can only be sold to private persons at a price of about 2,000 yuan per square meter. It is no difficult task to infer the consequences of such a mismatch between incomes earned by men handpicked in a highly elitist further education system, with high responsibilities in their jobs, and the purchase price of such flats.

China's reform and opening policies have authorized Chinese companies to sell that part of their production which is above and beyond the specified percentage they are under obligation to deliver to the relevant government agency at a predetermined price, to other purchasers at a price agreed between themselves and those purchasers. The consequences have been twofold: first, Chinese companies are able to generate a cash flow that can be invested in their industrial projects, which is what the Chinese authorities had in mind; and second, a new category of merchants and intermediaries has been created who market products on this open market.

For several years, these merchants have been making substantial profits, which have caused their incomes to be completely out of proportion with top executives' remunerations. Eventually, such an income differential may well bring about a class structure with greater differences than in the west. Such differences, however, may also have adverse side effects on the operation of some key services in the Chinese economy. As has been pointed out above, for instance, the size of the Chinese territory means that today's road, rail and air transport system is inadequate to meet actual requirements, and it is quite difficult to obtain seats on domestic flights and even on trains. Now, the emergence of this new class of merchants aggravates such difficulties: on the one hand, they have increased the demand for transport for their business activities, which involve travelling around; on the other hand, their income and financial power enable them to buy plane and train tickets on the black market at a price substantially higher than the official price. Inconveniencies caused by this situation may be a deterrent for foreign businessmen contemplating investment in China.

Finally, China must be careful not to upset the balance and consistency of its domestic political system. Westerners, because of their insufficient knowledge of China's history and of its present-day reality, which is the consequence of that history, are apt to analyze changes in China's domestic politics in the light of political patterns suggested by the western way of thinking, particularly of the democratic model developed in Greece at the very time when Confucius was putting together the elements of Chinese thought which had started to germinate under the Zhou. Although democracy is the political point of reference of all modern western countries - with the exception of Britain, which has retained the formal aspects of the monarchy and aristocracy - it is completely foreign to Chinese thought.

The republic in 1911 under Sun Yat Sen was merely the consequence of the Chinese people's determination to overturn the Qing dynasty. The republic was not an end, only a means. In 1949, the People's Republic of China was formed in accordance with a traditional Chinese pattern not in keeping with the model of western democracy. Any analysis of Chinese political reality made on the basis of the concept of western democracy is inadequate and is bound to lead to conclusions which will have adverse effects on China's political balance and coherence. The Chinese students who desire progress and prosperity for their country - generous and enthusiastic as they are - often lack sufficient experience and depth of thought, and when they advocate - passionately - the model they have known and seen to be working in the west, they repeat the mistake the Chinese republicans made in 1911. Unless China wants to expose itself to unrest and failure, it must not turn away from its history of four thousand years - the oldest in the world.

Westerners who would like to establish balanced, smooth and durable cooperation with China, must necessarily take the following four factors into account in their approach: the size of the country, its population, the specific characteristics of Chinese civilization, and the specific characteristics of its political system. An analysis of China's specific features shows that there are two mistakes that must be avoided: first, trying to impose on China models which have been developed and applied - sometimes with questionable success - in western countries; second, looking for similarities in regions or countries that would appear to have points in common with China.

Oversimplifying, some classify China with Eastern bloc countries, due to the purported similarity of their political systems, or, alternatively, with developing countries. China, however, clearly has its own political system; political systems along purely western lines, such as democracy, have not succeeded in China, as was amply proved by the failure of the 1911 republic and by the elimination of the Chinese communist party in Shanghai in 1927. Even if China dubs itself "communist" or "socialist", it always adds the qualification of "the Chinese way"; indeed, the part that corresponds to this "Chinese style" is in reality much more important than the formal part taken from western ideology. Like westerners today, the Christian missionaries - Roman Catholics as well as Protestants, but particularly the Jesuits, who were active in China in the 18th century - were wrong to believe that they would succeed in converting the Chinese to Christianity. And just as China could not become Christian then, it cannot become communist or marxist today. Its roots are elsewhere.

Those who classify China among the developing countries, including western managers who put their employees dealing with African or South American countries in charge of Chinese relations, too, make an even greater mistake. The consequences of such errors sometimes clearly appear in the economic competition between major western industrial groups on the Chinese market.

Finally, while the knowledge and experience acquired in Taiwan, Hong Kong, Singapore, Japan and Korea is extremely useful in a potentially successful approach to the Chinese market - on account of, at least, a partial identity of their and China's way of thinking and cultures - we must not rush into an ungrounded generalization. Taiwan is a Chinese province with a population of about 18 million, Hong Kong is a Chinese territory with a population of about 5 million, and the population of the state of Singapore, in which the Chinese account for a proportion of approximately 76%, is below 3 million. Regardless of the political issues between mainland China and Taiwan, the problems facing a country with over 1 billion inhabitants scattered throughout a country 3,400 miles long and 3,100 miles wide, are radically different from those other countries in the Chinese world have to solve. Japan and Korea, for instance, differ greatly from China, not only in their size and population, but also in their social and political situations.

There is no escape from the obligation to study, to learn in depth, China's specific character.

VARIOUS FORMS OF COOPERATION

At this point, let us summarize the conclusions of our various analyses so far. First of all, and contrary to what the Chinese often say, China's technology gap is not the main obstacle to the development of the country's economy. In many high-tech areas, China is in the middle section of the most modern countries; sometimes, it is even in the lead. It has both a further education system and the human resources necessary to catch up with the other countries within a fairly short period of time in the sectors where it is lagging behind. It has shown its ability in nuclear energy and in aerospace.

To try and solve the problems the Chinese economy must overcome only by means of technology transfer, is a wrong approach costing vast amounts of foreign currency without achieving the expected beneficial effects for the Chinese economy. The relative weakness of the Chinese economy results, on the one hand, from Chinese history, over which no one at home or abroad can have any control and, on the other hand, from a use of existing technology that is neither efficient nor profitable.

Until now, China has mainly had to buy goods, equipment and products abroad. The best-known major contracts for purchases of equipment relate to diesel generators, aircraft, the Daya Bay nuclear plant, the construction of the Shanghai underground system, and the Er Tan hydroelectric dam. The cost of these projects, however, only represents a smallish proportion of China's US$ 33 billion overall imports. China also imports, among other things, aluminium (China's annual aluminium output is about 500,000 for an overall consumption of 1.1 million tons), metal products, chemicals, etc.

The Chinese economy has been making rapid strides since 1979; in certain years, its growth in terms of volume exceeded 10%. This expansion was mainly bolstered by an increase of output in consumer goods.

The bulk of Chinese and foreign investment in China over the last few years was in manufacturing industries. Demand for energy, raw materials and semi-finished products rose correspondingly, but Chinese production was unable to keep pace. Consequently, imports of such products increased, which caused an imbalance in China's trade balance. As China has a limited supply of foreign currency, there was a shortage of these products on the domestic market.

The creation within the economic reformation process of an open market - on which, as mentioned above, Chinese companies may sell part of their products at a price which is freely negotiated with the buyer and which is based on market forces - is continuing to raise domestic prices to levels substantially above world prices. Thus foreign industrialists who have formed joint ventures with Chinese corporations to manufacture products which are partially meant to be exported in order to generate the currencies required to repay the investment, are penalized today because their cost prices are substantially above that of their foreign competitors. In effect, they have to buy the raw materials or semi-finished goods at the open Chinese market prices, which are higher than the world price their competitors have to pay elsewhere.

This situation is not without hope, at least in the long run, since China has sizeable mining resources; in some cases, the world's largest. These reserves are also very varied and encompass the whole range of commodities China requires to expand its economy. But what it is lacking, today, is the capital necessary to carry out its mining activities, and to build ore processing, refining, and metal transformation plants. For instance, China has huge coal deposits in various locations throughout its territory, but the financial means to step up coal mining are not there. Other coals mines are not worked because the Chinese railway system is not adequate to the country's transport requirements; indeed, many vital railway lines are still single track.

Then again, China has a very high potential in the field of energy generation. Chinese rivers permit the construction of large and numerous dams to generate hydroelectric power. Even if it is taken into consideration that the main waterfalls are located in areas where electricity consumption is low - the autonomous areas of Tibet and Guang Xi, or Guizhou province - power can be transported, without too much difficulty, to the main consumption areas, such as Guang Dong province, where industries are highly concentrated around Canton and Hong Kong, both of which can use power supplied by the Guang Xi autonomous area and by Guizhou province. As this province also has substantial bauxite reserves, it is possible to build aluminium plants to scale up output capacity.

China's huge population does not only provide a number of difficult problems - feeding, clothing, transport, accommodation, education and training - but is also a vast manpower reservoir and, at the same time, the world's largest market. As mentioned above, rural areas still represent the vast majority of the population, with nearly 80% of the total. If we except towns or provinces such as Shanghai or Liaoning respectively, which because of a foreign presence during the first half of the 20th century, have acquired an industrial tradition and have both qualified labor and experienced supervisors, most of the Chinese regions have a workforce of high quality. Since they come from rural areas, however, and thus form the first generation of industrial personnel, they are still short of experience, as well as of an industrial culture and background.

Foremen whose qualifications and skills are passed on and added to from one generation to the next, are not available in sufficient numbers, and their experience is too limited for them to be able adequately and efficiently to supervise teams of workers on production lines and in maintenance. Nonetheless, China's human resources are a vast potential. Once Chinese labor is adequately trained, they are extremely reliable. French construction groups which carried out major projects in China, have acknowledged the quality of Chinese labor. In some specialist areas - such as the erection of scaffolding, work with reinforced concrete, or welding - these firms observed that Chinese skilled workers were as competent and efficient as, and sometimes even more efficient than, the western instructors placed on the sites.

Managers of western companies which have formed industrial joint ventures in China all recognize that, at the end of their training, workers are capable of doing the jobs assigned to them, and engineers are
competent; shortcomings, if they exist, are most observed among supervisors. This is probably due both to the influence of the cultural
revolution and to the fact that foremen have only recently left their
rural background.

This list of strong, and potentially strong, points of the Chinese
economy, as well as of its weak points, provides an indication as to
how the rest of the world can help to meet China's needs and to generate profits and advantages for the country. China's three major requirements - emphasized in this contribution - are linked with the
shortage of capital, the lack of experience in both the management of
companies and in the country's economy at large, and the sometimes inadequate experience of middle management. China's three major assets
which are apt to attract foreign investors and businessmen are its
enormous reserves in raw materials and energy, the volume and quality
of Chinese labor, and the size of the Chinese market, even if the majority of the Chinese population are not yet solvent because of their
still very low purchasing power. China must, however, gain full control of three major factors in order to be able to carry through the
policy of reforming its economy and of opening up to the outside
world: its size and the resulting centrifugal forces, the growth of
its population, and its social and political transformation.

Potential foreign partners, too, must adhere to three fundamental pre-
cepts if they want to develop balanced, smooth, and durable relations
with China: they will have to speak the language fluently, become familiar with Chinese culture, and develop a reasonable understanding of
Chinese reality.

A FEW COMMENTS ABOUT COOPERATION

As mentioned above, the Confucian thoughts underlying the development
of China's culture and civilization for over four thousand years are
completely different from, and have other roots than, the Greek and

Judaeo-Christian outlook of the west. Westerners and the Chinese view each other's culture through the looking glass of their own respective philosophies. As a consequence, they cannot fully understand each other's outlook and way of thinking.

If we adopt the idea, generally accepted today, that the countries of the Pacific Rim and the countries with a Chinese culture background - to use Leon Van der Meersch's expression - are those that will engineer the strongest economic development over the decades to come, then countries such as West Germany or France, or supranational institutions like the EC, whose size is comparable to China's, must necessarily show an interest in China. China, after all, is the heavyweight of that area, and privileged relations with it may prove to be important. Apart from purely economic considerations, there are geo-political reasons for the development of such relations; but this is not the place to elaborate on such aspects. To emphasize technology transfer, as is done very frequently, is usually a mistake and always proves to be an inadequate approach as this is evidently not at all the area where China has the greatest needs.

Foreign businessmen who, through oversimplification or to satisfy their own short-term interests, will only regard China as a market where they can sell their equipment and technologies, condemn themselves to having merely superficial, fragile and short-lived relations with China. Those who, on the contrary, take into account a careful and thorough analysis of Chinese society and of the needs of the Chinese economy, opt for an approach taking into consideration China's own interests, and are prepared to give all the advantages that can be expected from real economic cooperation, will lay the basis for deep, strong and durable relations.

To achieve this, cooperation must be such that it gives the Chinese what they are really lacking, i.e. capital, experience in the management of new technologies, and training for technicians and workers, and for Chinese supervisors above all. Foreign contributions in these various areas must not be piecemeal but, on the contrary, form a coherent and consistent whole.

It is currently difficult to imagine any foreign investor putting any capital into China unless he is satisfied with the quality of the management of the company in which his investment is to be made. And this is, indeed, a vital conclusion: the ability to manage companies properly is what China is lacking most today.

The best approach therefore seems to be to choose a project essential to China, which will use raw materials of Chinese origin, and to ensure that the products manufactured by such a plant will effectively meet the requirements of the Chinese market. Preference should be given to projects whose results will gradually replace Chinese imports. This will have the double advantage of enabling China to save on foreign currencies, while foreign investors can depreciate invested funds and repatriate dividends in transferable currencies.

Once the technical and commercial feasibility study is completed, one or more major international groups of world renown could team up with one or more Chinese organizations active in the same sector, to work on a project that may well appeal to foreign investors. With the protection of guarantees and assurances concerning their ownership of invested capital, the payment and transferability of dividends, such foreign investors would contribute a certain share of the requisite capital, while the remainder would be financed through facilities arranged by a group of major international Chinese and foreign banks.

Such an approach would appear to be best suited to the requirements of the Chinese economy, as well as to the legitimate expectations of those who are interested in the development of relations with China. It would enable China to find a lasting solution to its constant trade balance deficit, and to acquire the industrial infrastructure it still lacks. It would also speed up the training of China's workforce.

This approach would give foreign partners the possibility of being profoundly and lastingly integrated in the Chinese economic fabric, to gain the required experience and knowledge of Chinese reality, to feel at ease in this market, and to invest in new opportunities. The time has come now for foreign industrialists, and European industrialists in particular, to attune their Chinese strategies to the actual needs

of the Chinese market, and to switch from a strategy based on technology transfer to a strategy aiming at developing modern technologies
inside China itself.

CHANGING CHINESE THINKING ABOUT TECHNOLOGY TRANSFER

Ryusuke Ikegami
Chief Representative, Shanghai Office,
Japan-China Association on Economy and Trade

INTRODUCTION

The year 1978, when China made its decision to adopt on open policy to the outside world, marked a major change in the direction of the debate in that country on the subject of technology transfer. The works published on this subject since then can be roughly divided into two categories in terms of their character: at first, most of them aimed at popularizing the new open policy and emphasized the necessity for it, including the introduction of foreign technology, or discussed practical details such as what kind of technology should be selected for the time being and the way in which it should be introduced, but in 1982 works began to appear that dealt with the subject of technology transfer on the theoretical level from many different angles, those dealing with the future orientation of technology transfer surpassing the others in terms of both quantity and content. The present paper is a survey of the main such works for the purpose of considering recent changes in the Chinese discussion of the introduction of foreign technology, and determining what Chinese policy on this subject is likely to be during the Seventh Five-Year Plan period, which starts in 1986.

I. CHINESE THINKING ABOUT TECHNOLOGY TRANSFER BEFORE 1982

After the founding of the People's Republic of China, the Chinese Communist Party has continuingly adopted an active policy of introduction of foreign technology, the only exception being the period of the "Gang of Four" in 1975-76, and there were four waves of transfer of foreign technology: 1953-58, 1963-65, 1971-74 and 1978-79. There was some technology transfer in other years as well, but not much in comparison with those periods.

---

* With the permission of the Institute of Developing Economies, Tokyo

Europe-Asia-Pacific Studies in Economy and Technology
Leuenberger (Ed.) From Technology Transfer
to Technology Management in China
© Springer-Verlag Berlin Heidelberg 1990

With the adoption of the policy of opening China to the outside
world by the Chinese government in 1978, the decisive step was taken
toward a major importation of technology and capital. That was a
period of opposition to the two consecutive prededing policies, i.e.
that of the "Gang of Four" in 1975-76 of refusing to introduce
foreign technology, and that of the Hua Guofeng regime in 1978-79 of
introducing large plant technology, which was, needless to say, a
reaction to the policy of the "Gang of Four" - indeed an over-
reaction, considering China's foreign exchange situation at that
time, as symbolized by the incident in which the contract with the
Nippon Steel Corporation for the Shanghai Baoshan Iron and Steel
Complex was cancelled. The policy line in the period of the Sixth
Five-Year Plan (1981-85) concerning the transfer of foreign
technology was basically oriented toward the criticism and rejection
of the policy lines of the preceding periods.

Let us consider what the different policies adopted during those
waves had in common and how they differed according to the international situation that China found itself in.

First, let us consider the different categories of technology transfer.

1. Classification according to the form of technology
a) Technical information
b) Introduction of design documentation
c) Introduction of design documentation and main equipment
d) Introduction of design documentation and entire plants
e) Introduction of design documentation, equipment, machinery,
   operating techniques and management techniques
Design documentation can be sublassified into basic design and detail design, but we will not make that distinction here. Point e) is
what is known as technology transfer on a full "turn-key" basis, including both hardware and software.

2. Classification according to how advanced the technology is
a) The latest technology

b) Technology developed since the sixties

c) Technology developed before 1960

The difference between technology developed since the sixties and technology developed before is the difference between batch systems and automatic systems in processing industries, and the difference between the use of solid fuels and liquid fuels. It can also be considered whether or not electronic technology has been adopted for the nervous system of the technology in question.

3. Classification according to the economic criteria of the technology

a) Technology transfer in which priority is given to technology import substitution

b) Technology transfer in which the technology is imported again and again in large quantities with the same specifications

The points of these classifications that China's technology transfer has had in common throughout the period in question are the introduction of design documentation and entire plants (point d) of classification 1), the latest technology (point a) of classification 2) and technology transfer in which priority is given to technology import substitution (point a) of classification 3). It was consistently its policy to import the latest technology to serve as a model. The basic reasons for this were China's foreign exchange position and its pride.

In the third way of technology transfer, there were two changes. First, a shift from the policy of technology import substitution to a policy of giving priority to economic efficiency was taking place. This is represented by the fact that thirteen urea plants all with the same specifications were imported from Mitsui Toatsu Chemicals and Kellog Company. Second, payment for software was first recognized in 1972 in connection with the introduction of various kinds of equipment to the Wuhan Iron and Steel Corporation, with Nippon Steel Corporation as the general contractor. Up to and including the second wave of technology transfer (1963-65), the Chinese did not even consider paying for software.

The more recent change in the eighties was the adoption of policies of introducing not just hardware but software as well, and of introducing technology other than the latest technology. Such a policy change is in conformity with the shift in emphasis from the construction of new plants to the technological upgrading of existing plants.

This change does not mean that China has forsaken its policies of introducing the latest technology and of import substitution, which it has mantained consistently since the fifties. It will continue to uphold pursue them. Rather, it should be interpreted as a widening of options to include both the latest technolgy and the technology of the sixties, as well as both hardware and software, in order to cope with the new situation.

Let us now consider the salient features of Chinese thinking on technology transfer since 1982.

II. THE QUESTION OF SOCIOECONOMIC DEVELOPMENT STRATEGY AND THE QUESTION OF TECHNOLGY TRANSFER AS MOST RECENTLY CONSIDERED

Since about 1982 there has been a great deal of discussion in China on the subject of socioeconomic development strategy.

In one of the earlier papers published on this subject, "Guanyu wogu jingji fazhan zhanlüe wenti de renshi" (The perception of the question of China's socioeconomic development strategy) (13), Sun Shangqing identifies population, technology and resources as the three basic determining elements of development strategy. According to him, technology is already a major means of development of production power in the modern age when considered from the Marxist point of view, and furthermore there are three theoretical means of developing production power: (1) increasing labor power, (2) rationally organizing the different elements constituting production power, and (3) innovating and revolutionizing technology. He suggest that technological innovation or "technological revolution" is the best way of developing production power stresses the importance of technology.*

This attitude originates in Deng Xiaoping's address to the National Science Conference in March 1978, which overturned the argument of rejecting technology transfer that held sway in 1975-76. According to Deng, "Knowledge is power", and this assertion theorized the point that knowledge is an important element of production power.

Among the papers that have dealt with the conditions for determining China's own development strategy is Yu Guangyuan's "Guanyu zhongguo shehuizhuyi jingji, shehui fazhan zhanlüe wenti de kexue yanjiu" (Scientific research on the question of socioeconomic development strategy of China's socialism) (19)** the points out five features of China's circumstances that are determinant with respect to China's development strategy: (1) the fact that it has a population of 1 billion; (2) the fact that China can be divided into two parts in terms of population distribution and the state of economic development, i.e., eastern China and western China; (3) the pronounced interregional gap in economic and technological terms; (4) the fact that 80 per cent of China's population of 1 billion live in rural areas; and (5) the large overall technological gap between China and the advanced countries of the world. In his comments on this international technological gap, Yu indicates his strategic ideas about technology transfer in quoting Marx on Newton ("There is no telling how much time it took Newton to discover his law, but in the present day even a schoolboy can learn it by attending class for an hour.") and stating that because of this technological gap, China can develop its economy by making use of the latest technology of other countries.

---

* In another paper (14), Sun Shangqing clearly indentifies technolical transformation, i.e. what is referred to as technological innovation or "technological revolution", as being one of the tasks that should be given the higest priority by China.
** This paper is reported to have been the text of an address delivered in Hong Kong in 1982 (data unknown)

The background of the frequent publication of such papers on socio-economic development strategy has been the collapse of the policy line of "extensive expansion" of production (i.e. the policy line of achieving a quantitative expansion of production power by means of high investment) that was so strongly espoused after the expulsion of the "Gang of Four" in 1976. The low economic growth around 1982 as a result of the economic adjustments that were made in 1979 and the following years gave rise to the quest for a new policy line as well as for a new development strategy with technological innovation as its central force.

With the declaration at the Twelfth National People's Congress of the Chinese Communist Party in September 1982 of the strategic goal of quadrupling the total industrial and agricultural production figure of 1980 by the year 2000, development strategy came to be discussed, mainly in terms of the already established policy line of "intensive development" of the production capacity (i.e. the policy line of achieving a qualitative development of the production capacity by pursuing economic efficiency) and the concrete policy of the "technological transformation" of existing enterprises as the embodiment of this theoretical line.

For instance, in his paper, "Guanyu Jishu Jinbu wenti" (On the question of technological progress (10),* Ma Hong, after asserting that the future development of production will have to rely on technological progress per se on the one hand, and "technological transformation" of obsolescent enterprises on the other, states that with a shortage of energy, transport capacity and funds, production should be increased by raising socioeconomic efficiency, i.e. by enhancing the quality and grade of products through technological progress on the part of enterprises and by economizing on energy and input materials. He also propounds the idea of attaining at least 50 per cent of the planned increase in total industrial and agricultural production of more than two thousand billion yuan in the twenty-year period up to the year 2000 by means of technological

* This article is reported to have been written on September 29, 1982.

progress. In addition, in discussing the long-term planning of production and science and technology, he identifies the purpose of technology transfer as being the faster development of Chinese science and the development and strenghtening of the country's self-reliance capacity. Ma also stresses the need to save time and money in China's technological development by means technology transfer involving advanced technology compatible with conditions in China, with priority given to technology in the form of software.

In another paper, "Guanyu shixian zhanlüe mubiao de buzhou wenti" (The question of the procedure to be followed to achieve strategic goals) (11), he proposes the strategic goal for China's techno-logical development of equipping all Chinese mining and industrial enterprises with the technology that was prevalent in the econo-mically advanced countries of the world in the seventies and early eighties by the year 2000.*

Of the many papers published on the subject of Chinese development strategy, Huang Wen's "Jishu xuanze yu jingji fazhan" (Economic development and the choice of technology) (7) restricts itself to the discussion of technology per se. Huang explicitly presents the following ideas on the strategy of technological development: ad-vanced technology has to be adopted in the core sectors and fields of the national economy, but considering China's limited financial and physical resources and plentiful labor, an effort should be made to achieve a balance between producer goods and labor power, and the country as a whole must have a multistratal technological structure. In other words, sectors representative of the developmental orien-tation of technology-intensive production should undertake technolo-gy-intensive production, and sectors capable of easily digesting new technologies in which an increase in labor input directly results in

---

* Another representative article published after the setting of the strategic goals for the year 2000 is Sun Yanfang (15). This article advocates putting emphasis on agriculture in China's development strategy and, in discussing the manufacturing industry, identifies the "technological transformation" of existing enterprises as the means of achieving reproduction on a progressive scale He also asserts the appropriateness and the necessary conditions for this from a macroeconomic viewpoint. He does not, however, say anything about technological development or technology transfer.

increased production and an improvement of economic efficiency should, as fas as possible, undertake labor-intensive production, using appropriate technology in large quantities and complementing it with handicraft-type technology.

However, Huang Wen does not think that the fact of whether or not any particular technology is up-to-date should be the sole criterion for China's selection of technology. Rather, considering the various limiting conditions, he sees labor-intensive industry as the proper mainstay of Chinese industry. Furthermore, as for the choice of technology, he opts for the kind that will mitigate the contradiction between economic development and the depletion of natural resources, i.e. the kind that economizes on natural resources and substitutes them. As for an interregional strategy of technology development, he is in favor of making inland areas mores self-sufficient with regard to their capacity of supplying industrial products. For this purpose, coastal areas would serve as a source of technology transfer to inland areas; the hitherto espoused inland/coast dichotomy, marked by exchange of the industrial products of coastal areas for the raw materials of inland areas, would thus be replaced. More specifically, he proposes that the coastal areas continue to import advanced technology from abroad and in so doing pass the equipment that is replaced on to inland areas, so that there will be a systematic upgrading and change in the products of the coastal areas and a transfer of production of lower-grade products to inland areas.

A work that deals with the subject of the medium-term and long-term goals of Chinese technology transfer is "Guanyu wogu duiwai jingji huodong de fazhan zhanlüe" (The development strategy of Chinese foreign economic activity) (2) by Gao Dichen, Wang Baochen and Chen Jiaqin. The book discusses the individual strategies for commodity trade, the borrowing of funds, the import and export of labor and science and technology, and management experience exchanges. The authors identify the strategic goals of science and technology and management experience exchanges as follows: in the 1980s, the main strategy was importation, as far as the country was capable, of appropriate advanced, reliable technologies and the overall improvement of economic efficiency through the absorption, digestion, and dissemination of imported technologies. The first five years were mainly devoted to the introduction of technology (software) and management know-how in order to make it possible to promote "technological transformation", while the next five years were devoted to a continuous innovation of technology in an effort to establish an innovation cycle so that economic, scientific and technological development could be furthered. In the 1990s, the main goals will be a pronounced increase in the level of manufacturing technology, manufacturing processes, and management, and the attainment of a certain standing for Chinese technology exports. The book does not, however, substantiate the feasibility of such goals.

As we have seen above, the discussion by experts of socioeconomic development strategy in China, which has only just begun, has started with a debate on the determining elements of development strategy. By 1982, many papers had appeared on this subject that pinpointed technology as the most important of the elements involved in the formulation of development strategy.

III. THE AVAILABLE OPTION IN AN AGE OF NEW TECHNOLOGICAL REVOLUTION

Since 1984 works have been published of works in China on the "new technological revolution", the first of them being Ma Hong´s paper "Zhuazhu jihui yingji xin de jishu geming" (Seizing the opportunity for a new technological revolution) (12). According to Ma, this discussion in China of the "new technological revolution" has been triggered by a veritable flood of publications in the United States, Japan and Europe on the development and application of new technologies, including information technology (microelectronics, optical fiber technology, etc.), biotechnology, technology for the development of new materials and new energy, ocean development technology, etc.; these publications heralded a "revolutionary" development of social productive capacity and new changes in the life of society and individuals thanks to such new technologies; and the taking in of such technology by China would contribute to reducing its economic and technological lag to a considerable extent. The Chinese discussion of the "new technological revolution" is in a way part of the discussion of technology transfer strategies.

The Chinese publications on the "new technological revolution" can be divided into three main categories according to the specific aspects they treat: (1) the position that China should assume toward the situation in the advanced industrial countries with respect to the development and application of new technologies; (2) the points of emphasis in the use of new technologies by China; and (3) the schedule or program for implementation.

In his above-mentioned paper, Ma Hong cites three options that China has with respect to the position it can assume, i.e. (2) disregarding new technologies and new industries arising in the advanced industrial countries as a phenomenon completely irrelevant to present conditions in China, (2) pursuing such new technologies and industries indiscriminately, and (3) keeping an eye on the development trends of new technologies and making ample use of them according to China's needs and possibilities. Ma maintains that the third option is the one that China should take in line with its Marxist ideology. Furthermore, in his paper "Xin jishu geming he woguo gongye de jishu jinbu" (The new technological revolution and the technological progress of Chinese industry) (18), Yang Mu reminds his fellow countrymen of such slogans of the ill-fated "Great Leap Forward" of years back as "Catch up and overtake on all fronts!" and "Make our own complete (economic and technological) system!", and advises them to recognize the lessons of that miserable failure, while advocating a selective adoption of goals and points of emphasis. What these two papers have in common is that both of them espouse a selective application of new technologies, although Ma Hong presents his argument in rather vague terms.

On the other hand, Huan Xiang adamantly maintains in his paper, "Xin jishu geming yu woguo duice" (The new technological revolution and how China should react to it) (5), that China should set itself the strategic goal of attaining, by the year 2000, the level of industrial and agricultural production and science and technology achieved by the world's most advanced countries in the early eighties. Huan further demands that by the turn of the century, China should be level with the industrialized countries at least in some areas, and that it should refuse to consider any other course than an all-out effort to "catch up from behind". With the memory of the fiasco of the "Great Leap Forward" still in mind, however, he felt it necessary to qualify such "catching up" by taking into account China's present conditions and developmental needs, the necessity of an adequate planning and guidance, as well as an organizational and scientific foundation.

Next, let us consider two representative views regarding what points should be emphasized with regard to the use of new technology by China. In the above paper, Huan Xiang cites microelectronics (including computers) as the best prospect for China, and biotechnology as the second best. This is because, first, microelectronics, and particularly computers, unlike most other areas of new technology, which have not yet gone beyond the experimental stage, is already in wide use and is exerting a considerable influence. Second, if China is able to achieve a solid foundation in this area, it will prove to be a considerable advantage in "catching up" in other areas as well, and will provide a basis for the adjustment and transformation of China's traditional industries. Third, it should be noted that China already has a certain foundation in biotechnology.

Another view on this subject is presented by Yang Mu in the paper mentioned above. Like Huan Xiang, he is of the opinion that the emphasis should be laid principally on microelectronics technology, at least for the time being. Furthermore, he identifies four conditions which presently have a bearing on efforts to improve China's industrial technology: (1) a shortage of domestic and foreign capital; (2) mounting pressure for an improvement of the quality of industrial products and for lowering their prices in a situation where competition among enterprises is now allowed; (3) mounting pressure in terms of the rising production costs of industrial products on account of rising energy and raw material prices and rising wages as a result of policy adjustment, by the central government; and (4) the need to shift the emphasis in the structure

of exports from energy to industrial products in order to achieve an
expansion of exports. He thus also maintains that such conditions
provide a motivation to improve the quality of industrial products
and to reduce the consumption of energy and raw materials, and that
microelectronic technology, and particularly the applied technology
of microequipment, is tailor-made for such purposes. At the same
time, Yang Mu argues that it is necessary to emphasize one area of
new technology in preference to others, while admitting that the
emphasis could shift, in a few years, to another area or other areas
even though none can presently compete with microelectronics in
terms of the stimulation of the technological progress of Chinese
industry.*

As for the third aspect of the "new technological revolution", i.e.
the schedule or program for the implementation new technologies, the
paper mentioned above by Huan Xiang presents a sharply defined opinion. Huan gives the following analysis of the possibilities
regarding the acquisition and application of new technologies by
China. He forecasts the following trends in the economies of the
advanced Western countries, excluding Japan, for the next 10-20
years: (1) very small likelihood of a resurgence of fixed capital
investment and thus of large-scale replacement of plant and
equipment; (2) the prevention of a rapid transformation of outdated
technology by the risk of labor disputes; and (3) the continutation
of a situation whereby a rapid formation of extensive markets for
new technologies is impossible, owing to the contradiction in
capitalist economies, which promote progress in science and technology for reasons of competition, while preventing application of
new technologies to production for the same reasons, with prevention
outweighing promotion. He concludes that the market for the application of new technology will continue to be a buyer's market for
the most part, and that this will work to the advantage of China in
terms of its need to import new technology. As for Japan, he
expresses the view that, although it will have the benefit of more
favorable conditions than other advanced industrial countries in the

---

* Another paper that clearly identifies microelectronic technology as the new
  technology that should be emphasized is Yang Dexiang and Liu Tiemin (17).

short run, it will have to face a market that is predominantly a buyer's market like all the other nations. In the same paper, Huan Xiang touches on the subject of the cycle of development of science and technology, saying that in spite of the fact that, historically seen, the time it takes from a scientific and technological invention to its application has been reduced, not much time has yet elapsed since the commencement of research in a whole series of new technology, and most of these new technologies are still in the research or development stage. He therefore concludes that in view of the fact that China has a certain grounding with respect to new technology, it will be able to catch up with the international level even if it has a later start.

He proposes a ten-point plan for "catching up from behind". Of the ten points, three are worthy of note as representing new strategic ideas.

One such idea is that of constructing a Chinese version of "Silicon Valley" for the purpose of developing new technologies and promoting their application on the largest scale possible, this would constitute an integrated entity for research and development, design, for the popularization and production of the new technologies, as well as for marketing them both at home and abroad.

Another original idea suggested by him in connection with the possibilities of introducing new technologies is that, although the very latest technology cannot be bought, China should not have any trouble acquiring what is considered to be very up-to-date new technology in general. That is to say, in view of the tremendous speed of technological innovation in the world today, it is possible to purchase good, up-to-date new technology that has come out only very recently.

His third original idea is that of making up for the insufficient number of trained personnel in China in the field of science and technology by having foreign personnel come to China in large numbers, not just to work directly, but also to help train Chinese personnel so that they can join the effort to develop new technologies.

On the "new technological revolution", "Xin jishu geming yu woguo zai guoji jishu maoyi zhong de diwei" (The new technological revolution and China's position with respect to the international technology trade) (4) by Hu Jun and Zhang Bigshen, is a paper that treats the new technology transfer in strategic terms. It suggests four ways of improving China's disadvantageous position in the international technology trade: (1) the switch from traditional industry to industries that produce products based on the latest advanced precision technology in the large industrial cities of the coastal areas; (2) the switch from an emphasis on hardware to an emphasis on software in the structure of technology transfer, and the promotion of "digesting" the technology involved in such transfer, and of a "return of service" from it; (3) a further development of the "transit point" role of "special economic zones" and "economic and technological development zones" in the technology trade (particularly the utilization of the advantageous location of "economic and technological development zones" in the vicinity of large metropolitan areas for the selective introduction of advanced technology on the basis, for instance, of licenses from multinational enterprises, and for the promotion of technology exports on the basis of joint production, joint research, and in other ways); and (4) the development of China's own intellectual resources with a view to an expansion of software exports, particularly exports of computer software.

As we have seen, the "new technological revolution" is being discussed in China with a certain degree of concreteness. However, there has not been enough discussion of the relationship between China's technology transfer strategy and its domestic socioeconomic development strategy and the possibilities and methods of introducing new technologies, which means that such topics, too, will have to be treated in greater detail in the future.

IV. THE DISCUSSION OF THE ROLE OF "SPECIAL ECONOMIC ZONES" IN
    CHINA'S TECHNOLOGY TRANSFER

Since 1984 works have been published on the future of China's
special economic zones" (SEZs). That year saw the publication of
several papers dealing with the theoretical problems involved in
making SEZs still more open to the outside in the context of the
general further development of China's policy of opening its doors
at that time. One representative contribution is Xu Dixin's "Guanyu
jin yibu duiwai kaifang de ji ge wenti" (Some problems in opening
China's doors wider to the outside world) (16). This paper discusses
three main topics: (1) the establishment of enterprises in SEZs and
similar "open cities" that are based on 100 per cent investment by
foreign capital enterprises; and (3) the economic models of SEZs and
"open cities". The subject of technology transfer is touched on in
topic (2).

Xu expresses the view that some marketing in China of products produced on the basis of advanced technology furnished by foreign capital interests should be allowed. This would reinforce the effect
of an absorption of foreign capital and of the introduction of
foreign technology by SEZs and "open cities", and stimulate the
improvement of the quality of the products made by Chinese enterprises. In other words, he suggests that SEZs should be given
greater opportunities to introduce advanced technologies contrary to
the Chinese government's prevailing policy of refusing to recognize
any domestic marketing of the products manufactured in SEZs, so that
they would be able to earn foreign exchange.*

It can be surmised that Xu Dixin wrote this paper at the insistent
request of the Chinese authorities, who had made the decision to
open China's doors wider, and that, furthermore, the background of
that decision was a sense of crisis prompted by the rapid progress
made by the advanced countries in the development of new technologies: the same motive as that for encouraging the discussion of

---

* Another paper that discusses the "special economic zones" and "open cities"
  from the same viewpoint is Ji Chongwei (8).

the "new technological revolution" as considered in the preceding section. Accordingly, any discussion of the SEZs against such a background can be considered, in a way, to be a discussion of strategic methodology with respect to technology transfer.

The year 1985 saw frequent publication, of works dealing with the development strategy of "special economic zones". The arguments presented fall roughly into the following two classifications.

The first viewpoint is represented by Liu Guoguang's "Shenzhen tequ de fazhan zhanlüe mubiao" (Stragetic development goals of the Shenzhen Special Economic Zone) (9), which sparked off the subsequent discussion on the development strategy of SEZs. The technique employed by this paper in order to demonstrate the validity of the author's views, is a comparison of several main views concerning (1) the development orientation of SEZs, (2) the industrial structure, and (3) the type of industry that should be encouraged to join such zones. First, with respect to the orientation of SEZs, he maintains that they should have an external rather than an internal or a dual orientation, the reason being that the purposes of their establishment are the introduction of foreign capital and an expansion of exports, both of which are characterized by an outward orientation. As for the industrial structure of SEZs, Lin lists three choices - emphasis on agriculture, emphasis on manufacturing, and emphasis on trade - and opts for manufacturing, because only in that case can SEZs be expected to be able to fulfill their original mission of serving as "four-purpose receiving windows" for the introduction of (1) funds, (2) technology, (3) management techniques, and (4) know-how, as well as providing a basis for exports. Likewise, with respect to the type of industry to be encouraged in SEZs, he says that the natural choice is the technology and know-how intensive type rather than the labor-intensive type, if SEZs are to serve as "receiving windows". Although he primarily characterizes SEZs in this way, he nevertheless advocates a conjunction of manufacturing and trade and the construction of "comprehensive special economic zones" with a combination of financial, tourist, service, real-estate, agricultural, and other operations.*

The second viewpoint, concerning the development strategy of SEZs, is represented by Huang Fangyi's paper "Dangqian woguo yinjin he duiwai jingji maoyi de zhiyue yinsu he gaijin shexiang" (Present limitations on China's technology transfer and foreign trade, and a view as to what improvements can be made) (6), in which he maintains that now, as SEZs are in transition from the stage of foundation to the stage of maturity, is the time to clearly define their character and orientation, Huang discusses how both should be defined on the basis of what has so far been accomplished in terms of the introduction of foreign capital. With regard to the character of SEZs, he poses the question of whether they belong to the category of export processing zones or to that of scientific industrial zones, and concludes that there is no choice but to put them in the first category, since an analysis of what has been achieved so far in terms of an introduction of foreign capital and technology, shows that one can hardly expect much of them in terms of any absorption and introduction of advanced technology. As for the orientation of the SEZs, he offers a clear-cut choice between an "inward" strategic orientation of import substitution and an "outward" strategic orientation, that is to say, an export orientation, and concludes that one has to take the "outward" export strategic orientation for granted, considering that SEZs are essentially export processing zones. Therefore he sees the main purpose of SEZs to be earners of foreign exchange through exports, and places the elements of their industrial structure in the descending order of (1) trade, (2) manufacturing, and (3) agriculture, pointing that the large metropolises are better equipped than SEZs to serve as the receiving end of a technology transfer involving "sophisticated, precision and vanguard" technology.

Let us put the differences between these two views concerning the development strategy of SEZs as represented by the two cited papers into sharper focus. On the one hand, Liu Guoguang's paper, which

* Among other articles that take the same stand, there are He Fenghua (3), and Dai Yuanchen and Shen Liren (1).

fines an "outward" development orientation as having two aspects. On the one hand it mainly advocates the introduction of a manufacturing industry that is technology and know-how intensive, since such zones are meant to serve as "receiving windows", on the other hand, in spite of stressing such strategic goals, it nevertheless advocates the formation of "comprehensive special economic zones". By contrast, Huang Fangyi's paper defines an "outward" development orientation as being synonymous with an "export" development orientation, and maintains that SEZs should concentrate on earning foreign exchange instead of introducing advanced technology, and become specialized as export processing zones.*

In spite of the fact that SEZs are going concerns, their development strategy has not been defined. The reason is that they were originally established by socialist China as experimental bases for the introduction of funds and technology from capitalist countries, to be operated on a trial-and-error basis; and this policy has not yet changed. Nevertheless, despite trial and error, the introduction of advanced technolgy by SEZs hat not progressed as much as China expected. Hence the occurrence of the debate on the development strategy of such zones since 1985, a debate which has been specifically based on their present conditions. Thus neither of the papers mentioned in this section discusses the socioeconomic development strategy or the technology transfer strategy of China as a whole; rather, both limit themselves to the SEZs. Huang Fangyi's paper advocates the specialization of such zones as export processing zones for the purpose of earning foreign exchange, while disregarding them in connection with "domestic" socioeconomic development strategy. On the other hand, Liu Guoguang's paper assigns them the role of "receiving windows" for the introduction of technology into China, but fails to discuss how they relate to domestic development strategy. This is something that remains to be clarified.

---

* However, in addition to this debate, a development policy oriented toward an outwardlooking economy as espoused by Liu Guoguang was adopted for the Seventh Five-Year Plan period (1986-90) at the National Working Conference of Special Exonomic Zones held in late December 1985 and early January 1986, the items of such a policy being the stepped-up introduction of foreign funds and advanced technology, and the promotion of exports.

CONCLUSION

In the present paper, several topics of recent Chinese discussions
of technology transfer were selected, and the salient features of
strategic arguments on the subject presented in terms of papers
considered to be representative of them. All of the papers quoted
were in agreement on the following points:

(1) Technological innovation is the most important element in the
development of production.
(2) The introduction of the latest advanced technology should be accomplished by all available means.
(3) In introducing technology for technology-intensive industry,
priority should be given to technologies which China needs and can
absorb.

There are, however, points in the above arguments that outsiders
would not consider to be very well defined:

(1) What kinds of technology-intensive industries should be given
the highest priority in technology transfer? Will the technology
introduced be only the kind that serves as a model, the goal being
technology import substitution? Or will large quantities of
technlogy with the same specifications be imported for the sake of
economy?
(2) What will be the percentage breakdown in the technology transfer
between the most advanced technology and technologies for the transformation of existing enterprises?
(3) Is there any disagreement as to how the "special economic zones"
and "open cities" should be made to function as introducers of
technology and as its disseminators throughout the country?

As we have already noted, since the establishment of the People's
Republic of China, it has consistently striven to achieve the introduction of advanced technology and technology import substitution.
Since the early eighties, it has added the technological transformation policy line, and the introduction of software in keeping
with this policy line. During the first half of the Seventh Five-

Year Plan period, the emphasis can be expected to be placed on the technological transformation policy line and the introduction of software, but observers think it very likely that the policy of introducing of large-scale plants in large quantities will be revived by the nineties for the Eighth Five-Year Plan.

REFERENCES

1. DAY YUANCHEN and SHEN LIREN. "Guanyu shenhzen jingji tequ fazhan zhanlüe de tantao" (A study of the development of the Shenzhen Special Economic Zone), Renmin ribao, December 13, 1985.

2. GAO DICHEN; WANG BAOCHEN; and CHEN JIAQIN. "Guanyu wogu duwai jingji huodong de fazhan zhanlüe" (The development strategy of Chinese foreign economic activity). Caimao jingji, 1982, No. 5.

3. HE FENGHUA. "Shenzhen tequ fazhan waixiang xing jingji de wenti" (The question of outward economic development in the Shenzhen Special Economic Zone), Jingji ribao, September 16, 1985.

4. HU JUN and ZHAN BINGSHEN. "Xin jishu geming yu woguo zai guoji jishu maoyi zhong de diwei" (The new technological revolution and China's position with respect to international technology trade), Guoji maoyi, 1985, No. 6.

5. HUAN XIANG. "Xin jishu geming yu woguo duice" (The new techno-logical revolution, and how China should react to it), Zhongguo shehui kexue, 1984, No. 4.

6. HUANG FANGYI. "Dangqian wogu yinjin he duiwai jingji maoyi de zhiyue yinsu he gaijin shexiang" (Present limitations on China's technology transfer and foreign trade, and a view as to what im-provements can be made), Jingji yanjiu, 1985, No. 12.

7. HUANG WEN. "Jishu xuanze yu jingji fazhan" (Economic development and the choice of technology), Ganjiang jingji, 1982, No. 10.

8. JI CHONGWEI. "Woguo duiwai kaifang zhengce de lilun he shijian" (The theory and practice of China's open policy to the outside), Jingji yanjiu, 1984, No. 11.

9. LIU GUOGUANG. "Shenzhen tequ de fahzan zhanlüe mubiao" (Strate-gic development goals of the Shenzhen Special Economic Zone), Renmin ribao (foreign edition), August 12, 1985.

10. MA HONG. "Guanyu jishu jinbu wenti" (On the question of techno-
    logical progress), in Tansuo jingji jianshe zhi lu (In quest of
    economic construction) (Shanghai: Shanghairenmin-chubanshe,
    1984).

11. -------. "Guanyu shixian zhanlüe mubiao de buzhou wenti" (The
    question of the procedure to be followed for achievement of
    strategic goals), Renmin ribao, October 28, 1982.

12. -------. "Zhuazhu jihui, yingjie xin de jishu geming" (Seizing
    the opportunity for a new technological revolution), Hongqi,
    1984, No. 6.

13. SUN SHANGQING. "Guanyu woguo jingji fazhan zhanlüe wenti de
    renshi" (The perception of the question of China's socioeconomic
    development strategy), Renmin ribao, April 12, 1982.

14. -------. "Shilun jishu gaizao wenti" (Essay on the question of
    technological transformation), Jingji yanjiu, 1982, No. 2.

15. SUN YAFANG. "Ershi-nian fan liang fan bu jin you zhengji
    baozheng erqie you jishu jingji baozheng" (A four-fold increase
    in twenty years has not only political assurance but also tech-
    nological and economic assurance), Renmin ribao, November 19,
    1982.

16. XU DIXIN. "Guanyu jin yibu duiwai kaifang de ji ge wenti" (Some
    problems in opening China's doors wider to the outside world),
    Renmin ribao, May 22, 1984.

17. YANG DEXIANG and LIU TIEMIN. "Qiantan xin jishu geming yu woguo
    gongye qiye de jishu gaizao" (A brief discussion of the new
    technological revolution and technological transformation of
    China's industrial enterprises), Jingji guanli, 1985, No. 9.

18. YANG MU. "Xin jishu geming he woguo gongye de jishu jinbu" (The
    new technological revolution and technological progress of Chi-
    nese industry), Jingji yanjiu, 1984, No. 9.

19. YU GUANGYUAN. "Guanyu zhongguo shehuizhuyi jingji, shehui fazhan
    zhanlüe wenti de kexue yanjiu" (Scientific research on the
    question of socioeconomic development strategy of China's
    socialism), in Jinji-shehui fazhan zhanlüe (Socioeconomic deve-
    lopment strategy) (Beijing: Zhongguo-shehui-kexue-chubanshe,
    1984).

Part Two

Case Studies

THE DEVELOPMENT OF THE CHINESE STEEL INDUSTRY

Jacques A. Astier
Consultant
11 Passage Foubert, 75013 Paris, France

INTRODUCTION

At first sight, China's achievements in steel production are impressive, as we can see from the following statistics (Table 1; in million
tonnes of steel produced per year).

Table 1: Comparison of the evolution of steel production

Year	World	Developing Countries [*]	China
1950	192	3.5	0.7
1960	346	17	7.6
1970	595	41	17.8
1980	716	100	37.2
1988	778	163.5	59.2

[*] i.e. Latin America, Africa (except the Republic of South
    Africa) and Asia (except Japan)

Source: I.I.S.I. Statistics (Brussels)

This means that in 1988, the developing countries produced 21% of the
world's raw steel, with China representing 36% of the developing countries and 7.6% of total world production.

We reach a similar picture if we compare the main developing countries
as regards their steel production (Table 2; again in million tonnes
per year).

Europe-Asia-Pacific Studies in Economy and Technology
Leuenberger (Ed.) From Technology Transfer
to Technology Management in China
© Springer-Verlag Berlin Heidelberg 1990

Table 2: Comparison of the main developing countries' steel production

Total, developing countries	163.5
Subtotal, countries listed below	147.75
China	59.2
Brazil	24.7
Republic of Korea	19.2
India	14.3
Taiwan	8.5
Mexico	7.8
D.P.R. of Korea	6.75
Argentina	3.7
Venezuela	3.6

Source: I.I.S.I. Statistics (Brussels)

Such a development and such achievements deserve comments, however, and we shall concentrate on the three following questions:
- What is the background to this development, especially from the historical and geographical points of view?
- What is the structure of the Chinese steel industry, in particular the methods and processes of steel production, where are the plants located, and what are the products of this industry?
- Finally, how is the transfer made, or rather, how is new technology created?

At this point, we must emphasize the fact that this paper is based on a number of visits to China and, in particular, on important research conducted this year under the direction of Prof. Gilbert Etienne and the Modern Asia Research Center of Geneva, on a comparison of the Chinese and Indian steel industries, to be published shortly as Steel in China and India - a Comparative Study.

I - THE BACKGROUND TO THE DEVELOPMENT OF THE CHINESE STEEL INDUSTRY

This background can be appreciated when observed from the two points of view of geography, including raw-material resources, and history, including important demographic and political aspects.

From the geographical point of view, we can see the analogy of the large Chinese subcontinent with the USA and the USSR and, to a lesser degree, with India and Brazil, if the comparison is with other developing countries. The analogy is that they are all vast land masses with far less contact with the outside that such islands or peninsulas as Japan, Korea, Taiwan, many countries of South East Asia, and most countries of Western Europe. This has important consequences for the market, especially for the steel products dealt with in this paper, in that they have to be distributed inland by rail, road or, sometimes, inland waterways. It is different on "islands and peninsulas" where much of the transport is effected on the ocean and where there is a far greater degree of contact with world trade, i.e. import and export of steel products, than in the subcontinents.

A similar situation exists for raw materials. "Islands and peninsulas", particularly if they have some deficiency in raw-material quality or reserves, will turn to imports. As is well-known, this is the case with iron ore and coal in Japan, Korea, Taiwan and, as regards Europe, in Italy, the Netherlands, Belgium and increasingly in France, Germany and the United Kingdom.

On the other hand China, like the USA and the USSR, is one of the world's countries with the largest coal reserves, as we can see in Table 3, which puts China's resources into more or less the same category as those of the USA and the USSR. Practically all kinds of coal from anthracite to lignite exist in China, but the best coking coal, so important for the "classical" iron and steel industry, is mainly to be found in the North Central area, particularly in Shanxi Province. As we shall see later, this was an incentive for China to develop its steel industry along the same lines as Western Europe and the USA in the 19th century, and the USSR in the 20th.

Table 3: Coal resources in China and the world

	known reserves in billion tons*
USA	396
USSR	349
China	201
all other countries	381
total	1,327

* L'énergie: le compte à rebours. Rapport au Club
de Rome (1978)

Some more recent information that we collected during our last trip to China indicates still larger coal resources. The figure quoted was in excess of 1,000bn tons (1,438bn tons...). In any case, it is likely that China has roughly one third of the world's coal resources.

Given such a situation it is normal that China has become one of the world's three big coal producers: with about 900m tons/year, it ranks as number one, just ahead of the US (about 850m tons/year) and the USSR (about 800m tons/year). This also explains why the development of the Chinese steel industry shows many similarities with what happened in the US 50-70 years before, and in the USSR 40-50 years before.

On the other hand, this situation with regard to energy and in particular to coal, is quite different from that of a number of developing countries which lack coal, mainly coking coal, but are rich in oil and natural gas. These countries - Mexico, Venezuela and Argentina in Latin America; Algeria, Libya, Egypt and Nigeria in Africa; Saudi Arabia and the Emirate, Indonesia and Malaysia in Asia - have developed new processes of gas-based direct reduction. In China, too, large gas and oil reserves could lead to such iron and steel making developments, but up until now, this has gone no further than the R&D stage while, due to the coking coal resources, all Chinese plants are based on the so-called "classical" process of reduction and melting iron ore in blast furnaces.

As regards iron ore, the Chinese situation is less fortunate: there may be plenty of resources, but their quality is not the best in the world. The overall amount of resources is of some significance, however, as can be seen from Table 4.

Table 4: Iron ore resources in China and the world

	known reserves in billion tons*
USSR	60
USA	16.6
Brazil	15.8
Australia	15.4
China	9.1
all other countries	37.5
total	154.4

* P. Salesse, Minerais de fer et fondants. Les Techniques de l'ingénieur, Paris 1989

More recent studies estimate Chinese iron ore reserves at 50bn tons. It must, however, be emphasized that as far as we know, there are no very large high-grade iron ore deposits as in Brazil or Australia. On the other hand, Chinese iron ore reserves are pretty well dispersed around the country, which, as we shall see, results in possibilities of developing the iron industry virtually everywhere in China. In any case, the quality aspect must be emphasized in two ways:

- Most of the deposits are of a low iron grade, and the crude ore requires more or less complex processing to achieve a high iron grade concentrate. This is costly in terms of both investment and operation; logically, China has taken advantage of the low prices on the world market and has imported iron ore, particularly from Australia and Brazil. Moverover, China is starting a joint mining venture in

Australia, at the Chanar iron mine, with the Australian Hamersley
Company. Table 5 shows the evolution of Chinese iron ore production
and concentration in relation to imports of foreign ore.

- A number of deposits contain iron ore associated with other minerals
  which can be valuable, such as Vanadium, but raise problems in iron
  and steel making. This is even more complicated in the case of important amounts of titanium oxide, rare earth, fluorine, etc.

Table 5:   Chinese iron ore supply (in million tons)

Year	Crude ore produced in China*	Usable ore produced in China*	Imported iron ore
1950	-	2.22	-
1960	112.8	66.63	0.6
1970	64.2	41.2	0.7
1980	112.6	61.8	7.3
1987	161.4	81.2	12.1

* 1960 should be taken with great caution, as it was the time of the
  "Great Leap Forward"

These considerations are an indication of the structure of Chinese
steel production, which will be one of the subjects of the next chapter.

In we turn to an historical point of view, we have to emphasize the
sheer size of the population, as well as their ancient civilization,
including industrial aspects. Again, with regard to outputs and inputs
in the steel industry, Chinese civilization is inward-oriented. Without going into any historical detail, we must point out two important
facts: first, the absence of the Industrial Revolution, as it occurred
in all developing countries, in the 19th century, and second, the existence of an important level of civilization, including not only education but also some industrial experience. This experience may have
been limited but it was present around the whole country, most notably
on the coast, in the ports, along the rivers, and in the big cities.

Another point to be borne in mind is the size of the population of China, which can be appreciated from the data presented by Tables 6 and 7.

Table 6: Comparison of the population of the main regions of the world (in millions)

Year	Western industrial countries	Eastern industrial countries	Developing countries total	China	total
1950	564	268	1,693	557	2,524
1975	732	360	2,974	928	4,066
1990	800	406	4,036	1,128	5,242
2000	841	431	4,847	1,257	6,119
2025	891	486	6,818	1,469	8,195

Source: Nations Unies, Les perspectives d'avenir de la population mondiale (New York, 1982)

Table 7: Comparison of the population of the main developing areas (in millions)

Year	Africa	Latin America	India	China	other Asia*
1950	220	164	368	557	423
1975	406	322	619	928	780
1990	635	459	821	1,128	1,117
2000	853	566	961	1,257	1,370
2025	1,541	865	1,323	1,469	1,961

* apart from Japan (included in Western industrialized countries), China, and India

Source: cf. Table 6

These tables show that roughly one out of five inhabitants of our planet is Chinese, although this ratio could slightly decline as a result of the Chinese authorities' efforts in the field of birth control. Table 7 indicates that the populations of all other developing regions are growing at faster rates, which will pose greater problems for those areas.

Finally, the political point of view should not be overlooked, although a detailed account of it is outside the scope of this contribution. It must be borne in mind that the People's Republic of China has put great emphasis on both agricultural and industrial development. Periods of great effort, such as the time of the Great Leap Forward (1958-1960), and periods of great unrest, such as the time of the Cultural Revolution (1966-1976), had significant consequences for the Chinese steel industry. They can best be appreciated with the help of Table 8, where detailed statistics provide the following information:

- The impact of the Great Leap Forward between 1958 and 1960, when production increased a great deal, particularly of crude iron ore mining, pig iron and crude steel, but less of finished, rolled steel products. Also, the quality of at least part of this production was probably not up to world standards. Later, especially in the wake of the end of Sino-Soviet cooperation in 1960, there was an important decrease in steel production.

- The influence of the Cultural Revolution, starting in 1966. For a number of years until 1976, there were many political and economic problems and consequences affecting steel production.

Table 8: Production of crude iron ore, pig iron, crude steel and
finished rolled steel products (in million tons)

Year	crude iron ore	pig iron	crude steel	finished rolled products
1949	0.59	0.25	0.15	0.14
1950	2.35	0.98	0.61	0.41
1951	2.70	1.45	0.90	0.67
1952	4.29	1.93	1.35	1.13
1953	5.82	2.23	1.77	1.15
1954	7.23	3.11	2.23	1.18
1955	9.60	3.87	2.28	2.26
1956	15.48	4.83	4.47	3.27
1957	19.37	5.94	5.35	4.36
1958	75.00	13.59	8.00	6.20
1959	96.59	21.91	13.87	9.35
1960	112.79	27.16	18.66	11.75
1961	51.59	12.81	8.70	6.58
1962	25.78	8.05	6.67	4.69
1963	24.22	7.41	7.62	5.39
1964	26.74	9.02	9.64	6.97
1965	31.49	10.77	12.23	8.95
1966	39.29	13.34	15.32	10.51
1967	29.62	9.63	10.29	7.40
1968	26.79	8.57	9.04	6.87
1969	43.33	12.80	13.33	9.56
1970	64.22	17.06	17.79	12.23
1971	81.46	21.00	21.32	14.41
1972	84.60	23.55	23.38	15.61
1973	91.64	24.90	25.22	16.84
1974	86.81	20.62	21.12	14.67
1975	96.93	24.49	23.90	16.22
1976	89.70	22.33	20.46	14.66
1977	93.84	25.05	23.74	16.33
1978	117.79	34.79	31.78	22.08
1979	118.76	36.73	34.48	24.97
1980	112.58	38.02	37.12	27.16
1981	104.59	34.17	35.60	26.70
1982	107.32	35.55	37.16	29.02
1983	113.39	37.38	40.02	30.72
1984	126.71	40.01	43.48	33.72
1985	131.10	43.84	46.79	36.92
1986	149.45	49.94	51.90	40.20
1987	161.43	50.20	56.02	43.80
1988		51.00	59.22	47.00

Source: Ministry of Metallurgy (Beijing) Statistical Data

II - THE STRUCTURE OF THE CHINESE IRON AND STEEL INDUSTRY

The structure of this industry can be described by means of steel production methods and processes, the size of plants, and the products.

II - 1 METHODS

China is quite classical with regard to steel production methods, in that its steel industry relies extensively on the conventional process of reducing and melting iron ores in blast furnaces. This is the method shown on the left in Figure 1.

Figure 1: The main iron and steel production methods

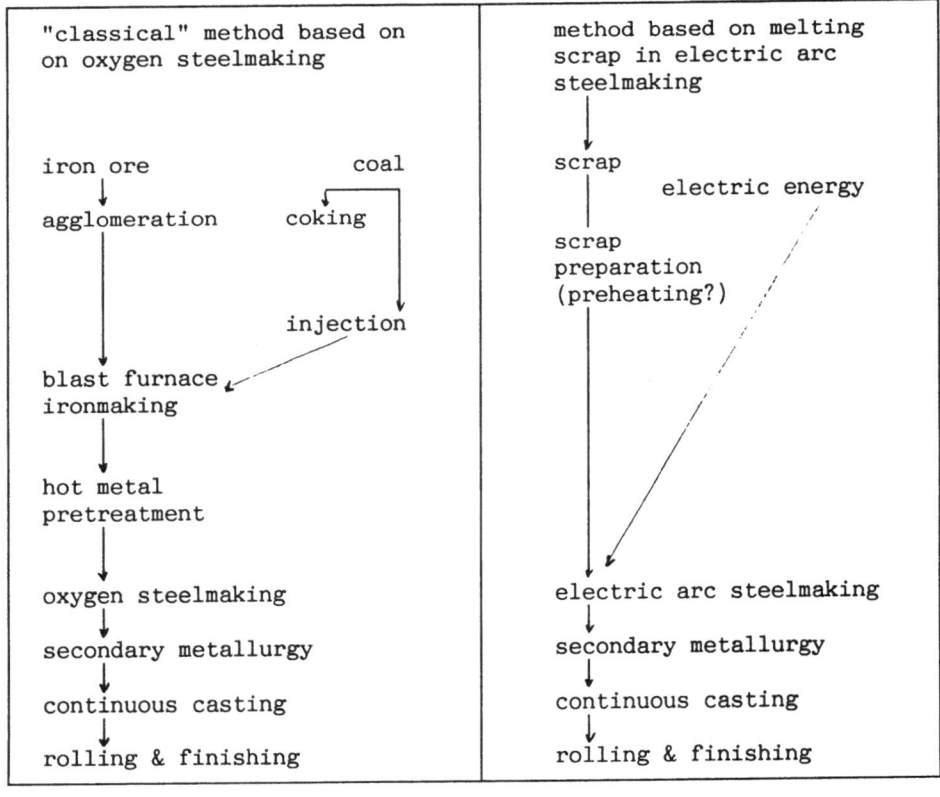

In the industrial countries, this method is used in large plants, usually with capacities of over 2 or million tons/year, and mainly for the production of flat products (sheets, plates) or some heavy or high-quality long products (rails, girders, wire rod). This method accounts for 50 to 70% of steel production in the industrialized countries.

The "semi-integrated" method on the right of Figure 1 is based on melting scrap in electric arc furnaces and is used in "mini-mills" whose capacity is between 200,000 and 1,000,000 tons/year. They chiefly produce light long products such as reinforcing rods, angles and other small or medium profiles. Such mills are in increasing competition with the large "classical" plants.

In the developing countries, such structures have been imitated or modified to a greater or lesser extent, since there were a number of problems:

- First, the size of the market. In a number of small countries with limited populations, there was clearly not much of a potential for integrated iron and steel plants, particularly large-size ones (unless they were modified, as we shall see later in the case of China).

- Second, the assumption that precisely this situation should provide mini-mills with an advantage. This, however, is not so simple since, in contrast with industrial countries, developing countries generally suffer from a lack of scrap and, frequently, from a weak power supply grid.

In China, the approach was generally quite different in that although there are some brand-new large integrated plants, such as Baoshan, and a number of mini-mills, most of the steel plants were started as small-scale classical integrated units. In other words, plants that were built in the 1950s or, in some cases, updated older plants, were quite similar to what had been developed in Western Europe and in the USA at the beginning of the "steel age", say, between 1870 and 1900. There was, however, the difference that a number of these small

Chinese plants developed fast, largely because it was known what had been achieved in the industrialized nations. Needless to say, some other plants followed a different evolution to become special steel plants; also, it is likely that some of these plants developed too fast at the time of the Great Leap Forward, only to fail later for some reason or other.

Returning to the second steel production method (on the right of Figure 1), which is usually referred to as "mini-mills", we find that there are some in China, but not so many as in most other countries. This is probably due to the following reasons:

- a short supply of electric energy
- a lack of scrap
- possibly some deficiencies in the construction of modern high-power electric arc furnaces and transformers

II - 2 THE MAJOR METALLURGICAL PROCESSES

Leaving aside the electric arc furnace (EAF) for the time being, the production of steel worldwide took place for the most part in open hearth furnaces. In the OECD countries, this process was later gradually replaced by oxygen-blown converters, which are known as basic oxygen furnaces (BOF). Figures 2 and 3 show the comparative evolution of open hearth and oxygen steelmaking in various countries including China.

It must be added at this point that in the 1950s and the early 1960s, there was an interim development in China of air-blown converters, particularly of the side-blown type: this was a very interesting technology, as we shall see in the next chapter on the transfer and creation of technology.

Figures 2 and 3 provide interesting indications:
- Especially in the 1950s and 1960s, China was not overly influenced by the USSR or the USA as regards these open hearth developments, certainly less than many other countries.

- China developed modern oxygen steelmaking not much later than some of the most highly developed countries such as Japan, or Western Europe.

Figure 2: The decline of open hearth steelmaking

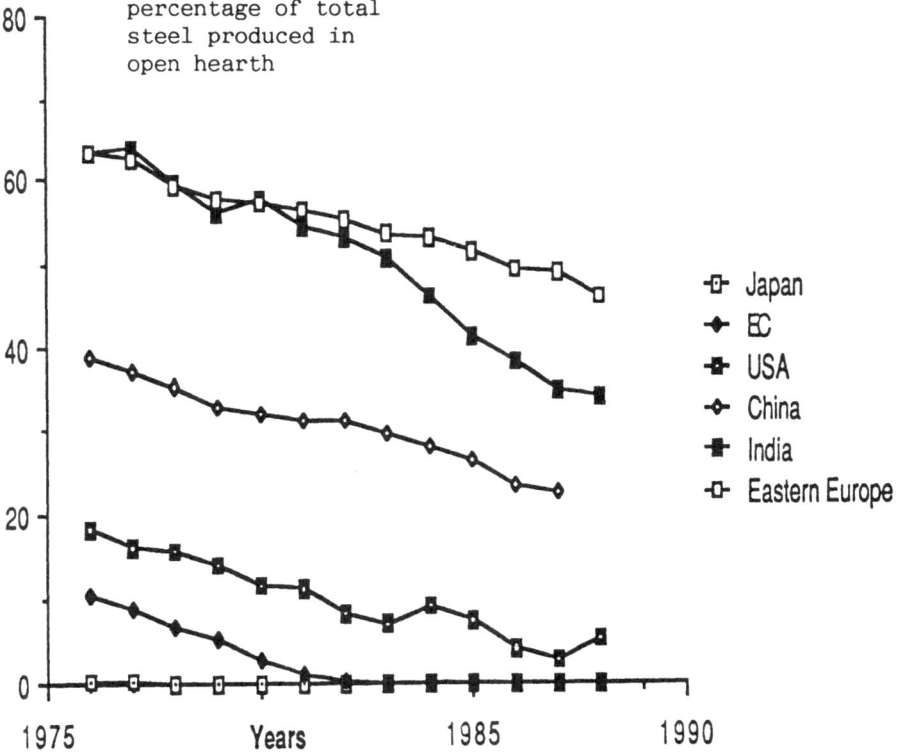

As far as ironmaking is concerned, Chinese metallurgy designed and built a large number of blast furnaces, starting from small-sized ones, and developed interesting technologies such as coal injection. Then again, they did not put a great deal of effort into new processes such as direct reduction or smelting reduction, except in the research and development phase. As mentioned previously, the large Chinese coal resources - mainly coking coal - may well explain, at least in part, why China kept developing the classic blast furnace.

Chinese developments in continuous casting will be dealt with in more detail in the last chapter.

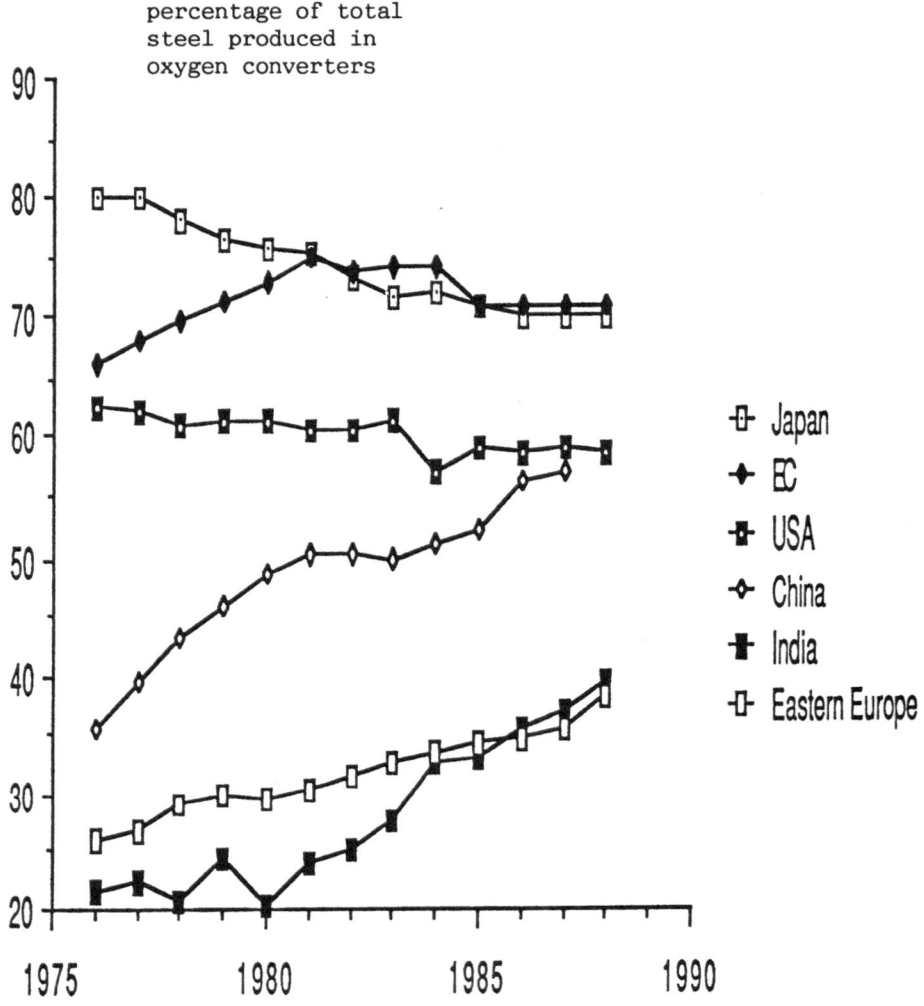

Figure 3: The evolution of oxygen steelmaking

percentage of total
steel produced in
oxygen converters

- ⊡ Japan
- ◆ EC
- ⊞ USA
- ◇ China
- ■ India
- ⊡ Eastern Europe

II - 3 PLANTS

It is remarkable to see that practically all the Chinese steel plants
have been built in or near cities, and frequently have quite a history. The difference here is that particularly in developing countries, steel plants are generally built "out in the sticks", i.e. often in desert or jungle areas or, at best, in agricultural surround-
ings. This takes us back to the historical consideration made in the
preceding chapter: that the Chinese steel industry has been built and

developed with the help of a labor force - including ordinary workers - that often had previous industrial experience. This is certainly one of the explanations of the fast development of this industry,
and as we shall see in the last chapter, it is probably connected with
the transfer and even creation of new technologies in the Chinese
steel industry.

As mentioned before, the "classical" steel production and plant-size
distribution structure of industrialized nations is as follows:

- 60-70% of the steel is produced in large integrated plants (i.e.
  with blast furnaces and oxygen steelmaking converters), usually with
  an output of between more than 2 million tons/year to 6-10 million
  tons/year.

- 30-40% of the steel is produced in "mini-mills" (i.e. based on scrap
  and electric arc furnace steelmaking). It must be noted that these
  "mini-mills" have been growing from initially small capacities of,
  say, 50,000-100,000 tons/year to the most frequent present-day capacities of between 150,000 and 1,000,000 tons/year.

In other words, industrial countries have two completely different populations of steel plants, differing according to

- size, i.e. annual steel production capacity;

- steel production methods;

- products, since the large integrated plants specialize in flat products (plates, sheets) and in heavy ones (rails, girders), as opposed to the light, long products (rods and merchant iron) constituting the main production of the mini-mills.

Most of the developing countries have been progressing along the same
lines, though with two imporant provisos:

- Due to market or financial problems, plants generally have smaller capacities, with large integrated plants operating in the 0.5-2 million tons/year range, while mini-mills become micro-mills at 30,000-50,000 tons/year.

- In some countries which have no coking coal but are rich in natural gas, the mini-mills have been integrated into plants based on iron ore directly reduced by gas; as in South Africa or India, such plants can also be based on iron ore directly reduced by non-coking coal.

As we have already noticed, China displays quite a different pattern, with:

- classical integrated plants with outputs ranging from 0.1 million tons/year (in some cases, even less) to 8 million tons/year;

- some mini-mills, but chiefly used to produce special-steel rather than producing the classical mini-mills product mix.

Again, we find a pattern which has some similarities with that of the European and US steel industry of the period between 1900 and 1950.

Something must also be said about the manpower and organization of steel plants in China: compared with industrialized countries (the USA, Western Europe, Japan), all the Chinese plants, with possibly a few exceptions such as Baoshan, employ vast numbers of personnel, bringing the total manpower to figures which are apt to be ten times that of industrial nations. This phenomenon is not peculiar to China but can also be found in the USSR, in India, and in some other developing countries. In the case of China, an analysis of this phenomenon results in the following three interesting observations:

- In the iron- and steel-production units proper (such as blast furnaces, steelmaking shops and rolling mills), the total workforce is not much different in number from that of corresponding units in industrial countries.

- The necessity of a substantial number of additional personnel can be attributed to a deficient infrastructure. In China, as in other developing countries, this requires a number of activities and services to be performed inside the steel industry which in industrial nations are supplied by other, external, private or public companies: the supply of electric energy, for example, water supply, port facilities, even the mining and processing of fluxes and refractories.

- Finally, many Chinese steel companies operate as "conglomerates", with a number of activities very far from steel production, such as township administration, hospitals, schools, various types of food production. We even found a steel company which had an important shipping subsidiary with several large vessels...

These conditions, seen as linked with the economic situation of a developing country, are interesting to note but will certainly change in the course of time. It is remarkable to see that the latest coastal Boashan plant is attempting to limit its activities to steelmaking and to increase its productivity, while aiming at a workforce total that could bear comparison with Japanese or Korean plants.

II - 4 STEEL PRODUCTS

In this context we have to remember that forty years ago, China was an "underdeveloped country" and, as a consequence, mainly needed relatively simply steel products, "commodities" such as concrete reinforcement rods, merchant iron for construction purposes, rails for the expansion of the railroad network, etc.

Such products were comparatively easy to produce and to roll in simple mills which were not too difficult to design and build in China, even if the first had to be imported, particularly from the USSR in the 1950s.

As a country develops its industrial civilization, the mix of required steel products usually becomes more complex, and China is no exception, needing increasing quantities of "special steels" in the sense of steel products designed for specialized uses - which is more general than the usual definition of "special steels". For our purposes, this means:

- higher-quality long products, such as wire rods or bars for specialized mechanical uses and the engineering industry;

- sheets and plates, again particularly for mechanical or construction engineering;

- steel pipes and, in particular, seamless pipes.

Today, China is increasingly in need of such products, for which it must rely on two sources:

- On the one hand, on imports, which have been running at a high level (see Table 9), turning China into the world's Number Two steel importer, just behind the USA! Some imports, especially of commodities, are only made because domestic production is still below requirements, but some of them concern special or quality steel not yet produced in China (or, again, not produced in sufficient quantity or quality).

- On the other hand, on production of increasingly sophisticated steel products. A number of plants, such as the new coastal Baoshan plant, as well as older plants like those in Shanghai, Tayuan, Wuhan, Anshan, etc., are moving in this direction.

Table 9: Semi-finished and finished steel consumption in China
(in million tons/year)

Year	Production	Imports	Apparent consumption*
1975	16.2	6.2	22.0
1980	27.1	5.0	31.7
1985	36.9	19.6	56.3
1986	40.2	17.4	57.4
1987	43.8	18.5	-
1988	47.0	-	-

* Exports are limited but are in any case taken into account.
  Source: I.I.S.I and Ministry of Metallurgy Statistics.

III - THE TRANSFER AND CREATION OF TECHNOLOGY

Looking at the operation of Chinese steel plants, as well as at their
evolution and updating, we find some very interesting characteristics.
Probably due to the prolonged isolation of Chinese metallurgists from
the rest of the world, we find that more technologies have been created, and that existing technologies function more efficiently, than
what tends to be the norm in developing countries.

III - 1 The concept of steel plant creation and expansion

The first aspect is the number of plants which have been created or,
more often, developed from old small enterprises, and the way they
were planned and specialized. The consequences of this aspect for the
structure of the Chinese steel industry has already been touched upon,
but it also has consequences for the creation of new technologies. If
the capacity of small plants is extended, then this also provides

opportunities to develop new technologies which may not be so easy to
include when large plants are built according to a rigid framework
with little room for change and evolution.

It must be emphasized again that the Chinese steel industry has constantly been developing from small or medium-sized plants, most of
them "home-designed" and "home-built", generally located near large
cities, where manpower was available both in quantity and quality.
This is in opposition to the concept whereby plants are designed in a
foreign country and built in the middle of nowhere as a "turnkey job"
from predominantly imported equipment. Once more, it must be borne in
mind that China is "walking on two feet" and has therefore also used
this concept when it was judged appropriate; this was the case with
Baoshan and, in the late 1950s, with Wuhan and Baotou. The latter two
plants are of Russian design while Baoshan is, as is well known, of
the most recent Japanese design.

III - 2 The selection of equipment

Another important aspect is the selection of equipment that has either
been designed in China, or imported. Two striking features emerge in
this context. First, very little equipment has been imported on a
turnkey basis. Such equipment is chiefly to be found in the large
modern coastal plant of Baoshan - although the expansion of this plant
is gradually being switched over to the Chinese manufacture of most of
the equipment - and in some large units like the hot strip mill and
the oxygen steelmaking shop of Wuhan. Second, the amount of second-
hand equipment bought in Europe, in the USA and in Japan: the Chinese
steel industry was in fact quickly aware of the amount of modern
equipment being phased out in those countries at the time of its own
restructuring, and did not miss this opportunity to acquire great
quantities of modern equipment in this way. Such equipment was generally dismantled by Chinese specialists, who later reassembled it at
home.

Nevertheless, it is astonishing to find the number of units, particularly experimental or industrial, but usually of small or medium size, which have been designed, engineered and built in China. A typical example can be found in the fast-developing field of continuous casting. Table 10 provides a summary of these developments:

- The total number of continuous casting units in China is impressive, although its increase is recent: out of 136 machines, 17 were started up in 1987, 30 in 1988, and 26 are scheduled to start in 1989. This amounts to a total of 74 units since the end of 1986!

- A total of 91 machines has been designed, engineered and built by the Chinese themselves.

Even if some of the machines may not be up to world standards, they still provide Chinese metallurgists with a great deal of experience, and facilitate the transfer and creation of technology.

Table 10:   Continuous casting in China and the world

	World number of units	Chinese number of units		
		total	imported	designed and built in China
Slab casters	371 (573 strands)	30 (33 strands)	10 25-300t capacity (13 strands)	20 13-90t capacity (10 strands)
Bloom casters	376 (1383 strands)	30 (100 strands)	7 (28 strands)	23 (72 strands)
Billet casters	704 (2374 strands)	76 (255 strands)	28 (115 strands)	48 (140 strands)

Source: Concast Standard. Continuous Casting Machines for Steel: World Survey 1989.

III - 3   The choice of processes

A third important point in this technology transfer (or creation) is
the choice of metallurgical processes.   In this respect, China again
differs somewhat from other, particularly developing, countries - and
the choice of steelmaking processes is a case in point. In the 1950s
and even in the 1960s, under the influence of the USA and the USSR,
most steelmaking plants were designed and built for the so-called
classical "open hearth process".   Under the aegis of the USA, the
following plants may be mentioned: Volta Redona (CSN) in Brazil, Altos
Hornos de Mexico and Fundidora in Mexico, Somisa in Argentina, as well
as Cornigliano in Italy. The USSR was involved in Bhilai in India, as
well as in Wuhan and Batou in China.

While such open hearth shops were built in Wuhan and Batou, however,
quite a number of small or medium-sized plants in China were trying to
produce steel without such expensive equipment. This was the important development of the side-blown converter, which displays some similarity with the Bessemer converter used in the industrial countries
some 70-80 years before.

This was very important for two reasons. First, it enabled Chinese
metallurgists to develop a cheap and efficient technology to produce
simple long products, which are very much in demand at that stage of
development. Second, it paved the way - as it did in the Europe of
the 1950s and 1960s - for a move from air-blown to oxygen-blown converters. It is worth noting that in China, the greatest part of this
move was made in the late 1960s and in the 1970s; the gap between
China and Western Europe was thus narrowed to 10-15 years!

There are, however, some limits and drawbacks: one of them was the
difficulty of designing and building modern oxygen plants in China,
another was the lacking ability to design and build large modern
steelmaking plants. This is only beginning to happen, but meanwhile,
many small-scale and efficient oxygen steel plants have been built and
are operating in China. Examples are provided in Table 11; the operation of oxygen steelmaking plants in China will be dealt with next.

Table 11:   Comparison of oxygen steelmaking operation

	Worldwide performance in large converters 200-300t units	Chinese performance	
		Baoshan 300t	Small units 15-25t
tap to tap time, mins.	30 - 40	35	23 - 40

III - 4   The operation of metallurgical units

The final point concerns the operation of metallurgical plants in China. In fact foreign training and technical assistance has been relatively limited. It was provided by the USSR at the time when Wuhan and Batou were being built, with Soviet assistance in a number of other plants, by Germany and Japan at Wuhan in the 1970s and, recently, by Japan again at Baoshan. If we compare the volume of this training and technical assistance (although this is difficult to quantify) with steel production, it would appear to be considerably lower than in Latin America, Africa or India. On the other hand, the gradual development in China of a number of small and medium-sized plants necessitated many Chinese engineers, technicians and operators training themselves in such a way that the performance of most of the Chinese iron and steel plants compares quite well with the best of world ratios. Blast furnace and oxygen steelmaking operations are cases in point.

With regard to blast furnace operation, we must first insist that Chinese metallurgists recognized the importance of the preparation of iron ore and the production of self-fluxing or basic sinter very early on (indeed, we may say that they did so before a number of industrialized nations...). Bearing in mind that an ideal blast furnace load per ton of pig iron would have to consist of

either approx. 1,600kg of self-fluxing or basic sinter

or   approx. 1,300-1,400kg of basic sinter and

200-300kg of high-quality lumps,

or, if starting from fine concentrates, of pellets instead of sinter (in the same quantity), the data provided by Table 12 show that China was not far from optimal values even 20 years ago. This quantity of sinter (or sinter + pellets), slightly above optimal values, means that Chinese iron ores are not the best, as indicated earlier on; and although great efforts were made to prepare them by concentration and mineral processing, the amount of gangue (and also ash in coal and coke) is on the high side, which leads to a slag volume somewhat above optimal values.

Even so, the performance of Chinese blast furnaces as summarized in Table 13 is quite good, both for production and productivity on the one hand, and for coke (or coke + coal injected in the tuyeres) on the other. It is worth noting that, although the small furnaces sometimes have too high a coke rate (due probably to low blast temperature, simple stoves and, possibly, bad coke), they show an excellent productivity and, with efficient small oxygen converters, can provide a good base for small integrated steel plants.

With regard to ironmaking technologies, we again find a typically Chinese mixture: they create new technologies such as coal injection, pioneering such a process worldwide, but if necessary transfer new technologies from abroad, such as bell-less charging systems or the construction of very large blast furnaces.

Table 12: The development of ore preparation in China: agglomerated iron ores / ton of pig iron

| Year | kg/t of pig iron | | |
	Sinter	Pellets	Total agglomerate
1950	194	347	541
1955	470	217	687
1960	864	77	941
1965	1,442	42	1,484
1970	1,396	91	1,487
1975	1,554	99	1,653
1980	1,516	63	1,579
1985	1,557	63	1,620
1987	1,588	86	1,674

Table 13:   Comparison of achievements in blast furnace operation 1985

	Worldwide	Chinese performance 1985 statistics	
Productivity t/m3/day	1.5 - 2.5	"key enterprises": medium/small-scale enterprises small local enterprises:	1.7 1.54 1.9
Coke rate kg/t pig iron (or coke + coal injection	450 - 500	average approx. small local enterprises	500 770

What applies to blast furnace steelmaking also applies to oxygen steelmaking: as Table 11 indicates, Chinese metallurgists have been building and operating very efficient small converters. For large units (200-300t converters) and for control systems, however, technology in the sense of both know-how and equipment has been acquired from abroad in various ways:

- new modern oxygen steelmaking shops, as in Wuhan or Baoshan,

- second-hand steelmaking shops, as in Shoudo,

- expert aid or technical assistance contracts.

Something must also be said about the other steelmaking process which is in competition with oxygen converters all over the world, but especially in developing countries: electric arc furnaces. Figure 4 indicates the relatively slow but continuous development of this process in the USA, Western Europe and Japan. In China, it appears more or less stationary in percentage terms; it is probably logical that China, like most developing countries, has problems in obtaining adequate supplies of

- scrap, particularly obsolete or capital scrap, which is always in short supply in a developing area,

- electric energy, due to the weakness of both the electric grid and power generation.

For this reason, China seems to have concentrated electric arc steel-making on the production of quality and special steel, mainly alloyed steel. The mass production of common steel, such as rods, in "mini-mills", has probably been postponed to some distant future.

Returning to continuous casting, the important Chinese efforts mentioned before (cf. Table 10) are leading to a development which could mean that in the future, up to about 50% of Chinese steel will be continuously cast by about 1993 or 1994. If we compare this to what has been achieved in industrial countries, as well as in a number of new industrial areas (cf. Figure 5), then we may come to the conclusion that on average, China lags about 10-15 years behind the industrialized nations.

Figure 4: The evolution of electric arc steelmaking

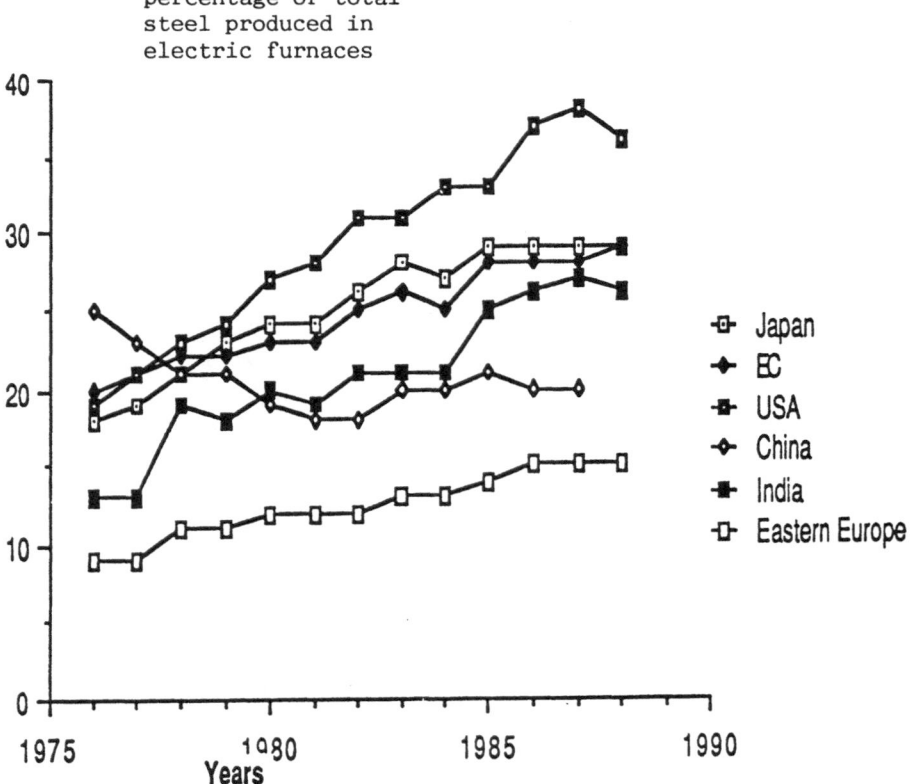

Finally, we have no precise statistical data - either for China or for industrial countries - on the modern trend to incorporate "secondary metallurgy" after the main steelmaking process and before continuous casting (cf. diagrams, Figure 1). Visits to China, however, have clearly shown important developments in this direction in many Chinese steelmaking shops.

Figure 5: The evolution of continuous casting in the main industrial countries and in China

% of total steel continuously cast

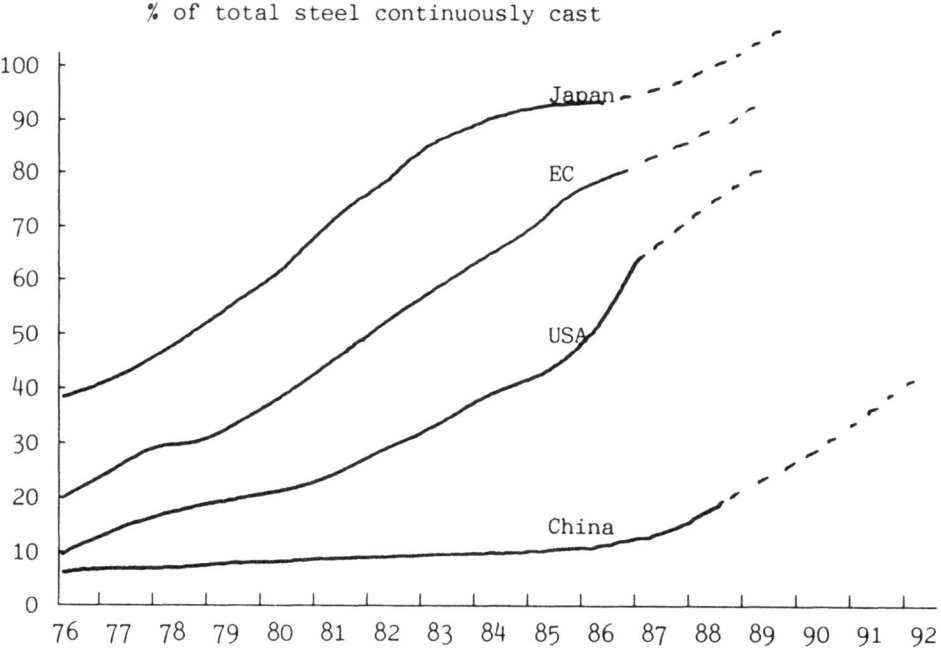

Source: OECD, The Role of Technology in Iron and Steel Developments (Paris 1989), and Ministry of Metallurgy (Beijing)
Statistics

CONCLUSIONS
‾‾‾‾‾‾‾‾‾‾‾

In our view, this brief survey of the evolution of the Chinese steel industry explains the remarkable results mentioned at the beginning of this paper. Let us remember that with 59 million tons of crude steel produced in 1988 (and probably 90-100 million tons by 2000), China ranks as the world's Number Four behind:

USSR   163.0 million tons (1988)
Japan   105.7 million tons (1988)
USA   90.1 million tons (1988)

or Number Five if we consider the European Community (approx. 137 million tons) as a unit. In China as elsewhere, however, there are a number of problems, some of which deserve special mention:

- as regards quantity: energy consumption per ton of finished steel products is still higher than in the older and later industrial countries;

- as regards quality: difficulties remain in making the increasingly better steel products required by an industrial civilization;

- as regards equipment: difficulties remain in designing, engineering and building modern large-scale metallurgical equipment and, in particular, control and automation systems for such large and high-productivity units;

- finally, in an industry where nothing is static, the need to adapt to changes in iron and steel making processes, as well as to the "near-shape" casting methods which are beginning to develop and will increase the importance of metallurgical studies.

Having said that, we consider the following points of our Chinese experience to be of major significance:

1) The decentralized efforts to build small or medium-sized classical plants, possibly 50 years behind the most advanced countries, but designed, built and operated by local people, and with an evolution proceeding as fast as possible to narrow this 50-year gap to about 5-10 years. The example of small-scale side-blown air converters modified to oxygen blowing is particularly striking.

2) Simultaneous efforts to remain informed - through various channels, such as technical assistance in some particularly critical fields, or by ordering the most modern equipment from abroad - about the most recent trends in world iron and steel technology, in order to develop all the plants as fast as new technologies can be absorbed.

3) It must be borne in mind that such an absorption of modern technology is limited both by the availability of sophisticated equipment and by the education and training available to everybody - managers, engineers, technicians, operating workers and maintenance staff.

As a summary, this highlights the relevance of a particular region's history and geography to industrial developments and technology transfer. More generally, it also emphasizes all the human aspects which unfortunately have often been neglected both in industrial and in developing countries. If industrial developments - say, in the steel industry - are planned from a purely technological point of view, the consequences may be quite detrimental: technology has an important human component, which must not be neglected.

TECHNOLOGY TRANSFER IN CHINA:
THE CASE OF OXYGEN-GENERATING EQUIPMENT IN STEEL INDUSTRY
1978-1988

Eduard B. Vermeer

CHINATEAM Consultants

P.O.Box 676, 2300AR Leiden, the Netherlands

INTRODUCTION

The production of steel requires large amounts of oxygen, some 50 to 60 cubic meters per ton. With the growth of China's steel industry, of about 8 percent per year during the last decade, its oxygen producing capacity has had to be expanded accordingly. In the following article, we will describe how this was effectuated through technology transfer, cooperation agreements, expansion of production and imports from foreign countries. Our interest focuses on

- the performance of the main producer of cryogenic equipment in China, the Hangzhou Oxygen Plant Manufactory, in its implemen-tation of a 10-year license agreement with the West German Company Linde A.G. concluded in 1978 with TECHIMPORT;

- the promotion by the national government of a coordinated development of native industries by means of technology transfer;

- the selection process of foreign oxygen-generating equipment in three Chinese steel plants under different administrative control and economic conditions, and their contract negotiations;

- the contract provisions for supervision of construction and training of operating personnel, and subsequent execution of the agreements.

We will try to draw some general lessons from these experiences, Chinese and for foreign enterprises. Finally, we will reflect on the

Europe-Asia-Pacific Studies in Economy and Technology
Leuenberger (Ed.) From Technology Transfer
to Technology Management in China
© Springer-Verlag Berlin Heidelberg 1990

present and future state of this branch of industry in China, in the light of the necessary expansion of steel making.

Technical background of oxygen production

The air separation technique was developed by Carl von Linde at the beginning of this century. Air under atmospheric pressure must be cooled to 81.5 Kelvin (minus 192 degrees Celsius), before condensation sets in; at 6 bar, this already occurs at 101 K, and the critical point of air is at 37.7 bar and 132.5 K. The necessary cooling down is achieved by expansion of filtered air which has been compressed (while being cooled to maintain its original temperature), and by using the reverse-flow heat exchanger. Oxygen becomes liquid at a higher temperature than nitrogen and argon. On the basis of this differential, air may be separated by rectification (that is, reverse-flow distillation) from nitrogen and other gases such as argon. Water, carbondioxide and carbonhydrates are removed from the air first, in a regenerator or reverse-flow heat exchanger, or with most modern installations in an molecular sieve-absorber. In the rectification column, the gas has a higher concentration of nitrogen while the liquid contains more oxygen. The required purities of oxygen and nitrogen are obtained by creating a reverse flow of liquid and gas successively in two rectification columns, between which a temperature difference of about 2.5 K is maintained by a pressure difference in the two columns (5.7 and 1.3 bar respectively, with the low-pressure technique of Linde equipment). The liquid oxygen and nitrogen are pumped into tanks, and the gaseous oxygen and nitrogen are withdrawn and compressed to the required densities. See Chart 1.

The air separation equipment has to meet high quality re-quirements. The material, originally copper but nowadays aluminum/manganate, must withstand extreme variations in temperature and the various parts of the installation must be optimized in order to achieve the required quantities and purities with a minimum use of energy.

1 Air compressor

Compression of air to approx. 6 bar in the multi-stage turbo-compressor with inter-mediate- and after-coolers.

2 Plate heat exchanger

(Reversing heat exchanger)
Cooling of air to liquefying temperature and deposition of water and carbon dioxide on the exchanger surfaces. Warming up of nitrogen-rich residual gas and removal of deposited water and carbon dioxide by residual gas. Both streams are reversed by a timer. Warming up of the oxygen and nitrogen products and expansion turbine air in separate passages.

3 Expansion turbine

Expansion and cooling of part of the air to cover cold losses.

4 Rectification column

Pre-separation of air into an oxygen-enriched bottom product and nitrogen as top product in the medium-pressure column 4 A. Final separation of the previous products into pure oxygen, pure nitrogen and nitrogen-rich residual gas in low-pressure column 4 B.

5 Adsorber

Purification of the liquids from hydrocarbons in particular acetylene, which have been contained in the process air. For this purpose the recycle adsorber 5 B is combined with a liquid oxygen pump.

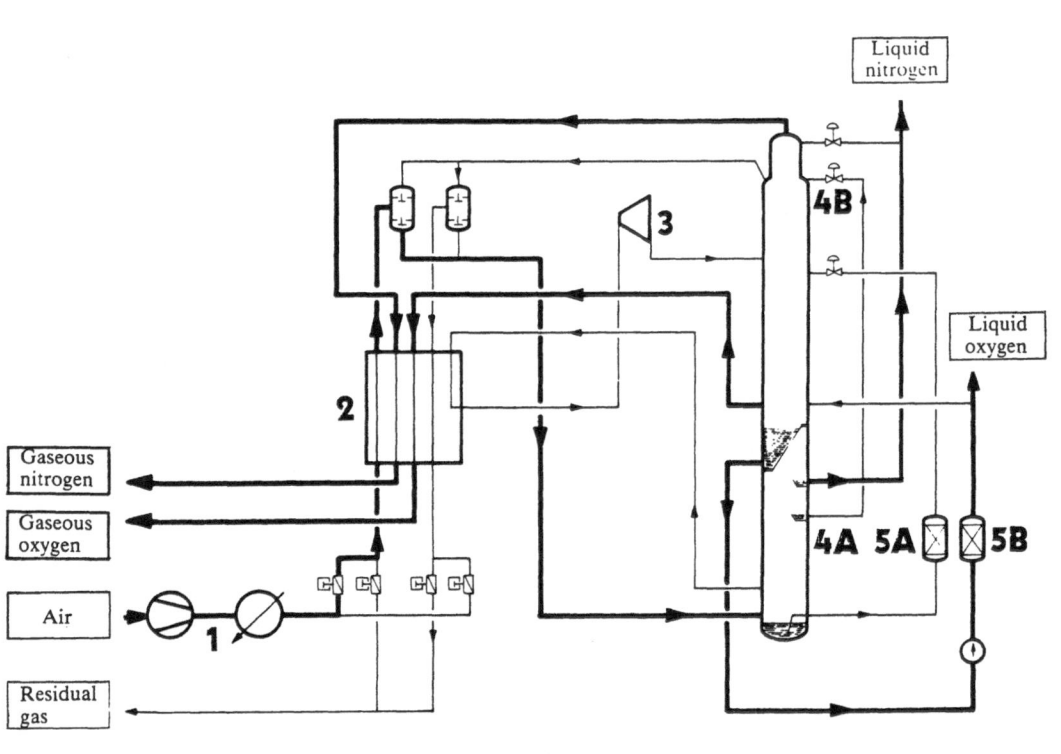

<u>Customer requirements</u>

Depending on customer needs, various amounts of oxygen, nitrogen, argon and rare gases may be produced. Steel factories want mainly oxygen (and may use argon as a protective gas), petrochemical factories want mainly nitrogen. The equipment is tailored to meet their respective quantity and purity requirements (liquid or gaseous). There are some economies of scale, both in the cost of equipment and installation and in operating costs. The air and oxygen compressors of the currently largest units of 30,000 cu.m./h oxygen use about 20 percent less electricity than the 6,000 cu.m./h models per unit product. For all users, reliability is a must. Most, but not all steel factories have liquid oxygen tanks containing a few hours' supply only. If the purity of oxygen drops (usually this is due to too heavy drawing), the quality of the steel will suffer.

FOREIGN AND CHINESE PRODUCERS, AND THE LINDE-HANGYANG
COOPERATION

<u>History of oxygen-manufacturing equipment in China</u>

In 1953, China started trial-production of air separation equipment in Hangzhou, first with only a small capacity (30 Nm3/h, followed by 50, 150 and 300 Nm3/h). Cooperation with the Soviet Union was realized in 1958, and Hangzhou started joint production of 3,350 Nm3/h units, with Russian advisers. Their designs at the time were still based on Linde's pre-war designs. In 1960, the air separation equipment industry was recognized by the Chinese government as a separate branch of the General Purpose Machinery under the Ministry of Machinery. In the early 1960s, China had to rely on itself, but by 1965, it approached Japanese firms to start joint production of larger plants. Although some large 5,000 - 6,000 Nm3/h sets (usually complete sets) were imported from Japan, France and West Germany in 1965 and 1966, and again in the early 1970s, no cooperation came about. Around 1966, following West German and Japanese examples, the industry changed over from copper to aluminum, and for its larger units the high-pressure process was abandoned in favour of the low-pressure process. By 1970, China produced about 20 sets of cryogenic equipment with capacities larger than 1,000 Nm3/h oxygen per year, mainly in sizes of 3,200-3,300 Nm3/h (the FL-3350-III and KFS-21,000 models) and later also with a capacity of 5,700 Nm3/h (the KFD-41,000 model). Until 1981, 255 of such sets had been produced. However, Hangzhou and the other factories produced mainly small-size cryogenic equipment: between 1953 and 1981, total output was 4,358 sets with a total capacity of 990,000 Nm3/h and a total weight of 219,000 tons. (Zhongguo Jingji Nianjian Bianjibu (ed.), Zhongguo Jixie Gongyedi Fazhan (Development of China's Machinery Industry), Beijing 1983, p. 88-89). Most of the appendages were made in China, but larger-size oxygen compressors, valves and meters were still imported.

During the early and mid-1970s, China had started to manufacture 6,000 Nm3/h oxygen producing units (the KFD-41,000 model) in Hangzhou, using designs which imitated foreign imported models. Designed purities were 99.6 % O2 and 99.99 % N2. A 400-page Chinese publication written in 1975 "Principles and Operation of Oxygen-manufacturing Equipment" (Zhiyangjidi Yuanli yu Caozuo, Beijing 1977) described the technical features of these models. Results, however, were unsatisfactory: neither in reliability and purity, nor in energy consumption could the foreign standards be reached. After very ambitious goals for a rapid expansion of the steel industry (with a doubling of output) had been adopted as part of the 1978-1985 economic development plan in 1978, the State Planning Commission decided that foreign cooperation was necessary in order to reduce China's dependence on imports. 1978 was the first year of the Open Policy, and in an expansionist mood the Chinese government started negotiating, and concluding, many contracts with foreign companies.

The Linde-Hangyang contract

In 1977, the China National Technical Import Corporation TECHIMPORT, a governmental organization under the Ministry of Foreign Economic Relations and Trade MOFERT, contacted Linde A.G. representatives (who were in Beijing for delivery contracts) and asked for a production license of Linde´s 10,000 Nm3/h equipment on behalf of the Hangzhou Oxygen Plant Manufactory. This came as a surprise to Linde.

Linde had been selected as a partner because it was considered to be the technological leader in its field, and the few installations it had sold to China (three 10,000 Nm3/h oxygen units, one of which with production of rare gases, sold to Taiyuan Iron and Steel Cy in 1965 and to Wuhan Iron and Steel Cy in 1972, and two plants for hydrodealkylation and concentration of hydrogen-rich gas sold to Yanshan Petrochemical General Corporation in Beijing in 1975) had been performing very well. Unlike the Japanese installations, those Linde had always reached or even surpassed the set standards. So no other Company besides Linde was considered.

When the Chinese approached Linde, their mind had already been set on a definitive size: a capacity of 10,000 Nm3/h. At the time, this was a high level, which went considerably beyond the 6,000 Nm3/h units already produced by the Hangzhou factory (called Hangyang for short), but not the largest size Linde could supply. Unknown to Linde, Hangyang had already tried to manufacture 10,000 Nm3/h units, but their results had been disappointing. (One set had actually been supplied to a steel factory, Anshan Iron and Steel Company; several years later, its performance was improved by upgrading). The Chinese engineers reckoned that, once the newest Linde techniques had been mastered, they would be able to adjust the size to other capacities. Linde was not asked for an opinion on the proposed size.

The officials of TECHIMPORT and of the Ministry of Machine Building Industry told Linde that China was considering a few dozen projects to be realized in the next decade, for which Linde could be a main supplier and subsupplier if it concluded a joint production contract with Hangyang. When negotiations started, the Chinese at first wanted to have all the know-how, but without paying a fee. The fees for the knowledge transfer, in their opinion, would be paid later by the end-user of the equipment. Linde, however, demanded a fixed sum for the design and production documents, with subsequent royalties for every unit produced together with Hangyang. Linde´s idea was, to put it simply, participation in order to sell. The Chinese demanded a cooperation period of ten years, without giving their reasons for that duration. At the back of their minds, they envisaged a protracted period of technical cooperation which would bring their design and production knowledge up to advanced world levels. As it happened, the expectations of neither party came true.

After discussions, in December 1978 both sides agreed on a 10-year contract, under which in the years 1979-1981 eight plants of 10,000 Nm3/h capacity would be built jointly on the basis of the Linde design. The level of Chinese participation would increase from 8 percent (of the Linde scope of supply) in money terms for the first four sets, to be produced in 1979, to 30 percent for the fifth and sixth sets, to be produced in 1980, and rise to 72 percent with the seventh and eight sets in 1981. Engineers and technicians from Hangyang would receive a total of 135 months of training in West Germany, and Linde engineers would stay in Hangzhou for supervisory work for two years - 1979 and 1980. There was no technical reason for the adoption of a three year period or for the number of eight plants, only a financial one. The technical negotiations between the Linde team (three high-powered executives) and the Chinese representatives of the Ministry and of Hangzhou Oxygen Plant Manufactory ran smoothly, and the commercial negotiations with TECHIMPORT did not last long either. From the Chinese side, there was a sense of urgency, and a recognition of the benefits cooperation with Linde would bring to the Hangzhou factory and China. The contract price and method of payment did not appear to be major concerns. However, a reduction was achieved by the Chinese by reducing the amount of training. It was agreed that Linde would receive 4.85 million DM for the license and documents, and about 400,000 DM for each of the eight sets produced. The contract also stipulated that Linde would supply certain materials and appendages such as air and oxygen compressors, which are not manufactured by Linde itself. However, Hangzhou could import these and other appendages directly from the producers.

Ten years later, both sides still felt that the contract price had been reasonable and profitable to both sides. But on some other scores both sides were dissatisfied, and this disappointment is still keenly felt today. Part of the problem lay in the ambiguities of the contract itself, and part in unrealistic Chinese expectations and mutual misunderstandings. The main exacerbating factor, however, was the Chinese governmental reaction to the economic crisis of 1981 and its subsequent retrenchment policies, which brought production of large oxygen-generating plants to a halt. In the contract, the license fee covered several distinct things: supply of documents, training in China and training in Europe. For financial reasons the Chinese side had insisted on reducing the amount of technical assistance in China from the originally pro-posed 15 months by more than one half. The number of trainees in Germany was fixed at 135 man-months. The Chinese did not want to make provisions for a subsequent check and follow-up of training results by Linde engineers - something strongly recommended by Linde.

Following the conclusion of the contract, a training plan was agreed upon early in 1979. All trainees had already been employed by the Hangzhou plant for many years, and carefully selected. Designers, computer specialists, mechanical engineers and aluminum welders went to Germany. One German senior engineer stayed as an overseer at the Hangzhou plant for two years, and went out of his way (and beyond the terms of the contract) to teach general problems of organization of production as well. The Germans noted that, with their help, what the factory produced was good quality, but when the German supervisors were on holidays, quality dropped. When the Germans left after two years, in accordance with the contract terms, they felt that the Chinese had acquired adequate knowledge and training to do a proper job, but they were not certain whether Hangyang's performance would be consistently up to the mark. Later, they had the impression (wrongly, as it appears) that many of the people trained in Germany were transferred to other companies, notably to the China National Air Separation Company established in 1982.

A special problem was posed by the computer software for design calculations. In Germany, four Chinese process engineers were trained for six months in making process calculations on Linde's IBM computers, and they were given examples of how to optimize such a process. Back in China, however, they found no computer which could accommodate this programme. The nearby Steam Turbine Factory (which was under the same ministry as the Hangzhou Factory) had recently imported a Siemens computer as part of a larger deal, and that was offered for use. The Linde people thereupon modified the programme, which required considerable effort, and it took a year before it could be implemented. Most likely, some of the know-how acquired by training during Germany was lost in the process. The lack of their own computer must have hampered design calculations at the Hangzhou plant. During the 1980s, several Chinese delegations which visited Linde for some weeks were introduced to the newest methods of process design calculations. Surprisingly, the German engineers rate the Hangzhou Factory process designers higher than the Chinese do themselves. They believe some five of the Hangyang engineers are quite capable of making the necessary design calculations for sizes between 10,000 and 30,000 Nm3/h on their own, but the Factory management does not think so. Hangyang still relies on the Steam Turbine Factory computer today, but now plans to acquire a British computer (Hewlett-Packard).

<u>After a boom, the crisis of 1980-1982</u>
was a top year for Linde in China. Under the license contract with Hangyang, three plants with installations for rare gases were immediately sold to the steel works of Wuhan and Taiyuan and one plant to the Daqing Petrochemical works. Apart from the Hangyang contract, the Company sold three 30,000 Nm3/h oxygen/nitrogen plants, three rectisol wash units and three NH3 synthesis gas plants (all in cooperation with Ube Industries, Tokyo). Then the crisis intervened, and it was to take four years before the next installations would be contracted for, in 1983. The four plants contracted for with the Hangzhou Company in 1979 started up only 5 to 6 years (1985, in the case of Daqing) later, which was twice the ordinary lead time (one year for design, production and installation each). Relations with Hangyang were practically severed.

While the 1980 cancellation of many foreign contracts by the Chinese government is well known and remembered in the West, the subsequent crisis in China's heavy industry has often been over-looked. Under the Three-year Readjustment Plan 1979-1981 industrial growth, particularly of nationally-owned heavy industries, slowed down. In 1981, iron production decreased by 10 percent and steel production

by 4 percent (Zhongguo Tongjiju (ed.), Zhongguo Gongye Jingji Tongji
Ziliao 1949-1984 (Statistical Materials on China's Industrial
Economy), Beijing 1985, p.50). By itself, this one time dip may not
seem too serious, as the next year, steel output regained the 1980
level. However, the Government had reacted to (and had provoked) the
crisis by severely curbing its long-term investment plans for the
steel industry. Such investments dropped to RMB 3 billion in 1980
and RMB 2.5 billion in 1981, and then regained a level of about RMB
3.5 billion in the years 1982-1984 (Zhongguo 1986 Gangtie Gongye
Nianjian (China 1986 Iron and Steel Industry Yearbook), Beijing
1986). The machinery industry was completely overhauled, and
reoriented towards serving consumer industries rather than heavy
industry. In 1981, the government constituted "readjustment groups"
for 17 different sections of the machinery industry, each of which
was assigned the task of drawing up nationwide readjustment plans,
which would promote cooperation between the enterprises within each
section. The plans should devise technological policies, seek to
avoid unnecessary duplications and stop unnecessary imports (Nanjing
Daxue Dilixi Jingji Dili Jiaoyanshi (ed.), Zhongguo Jingji Dili
Cankao Ziliao (Reference Materials on China's Economic Geography),
n.p.n.d. p.430-1 and Zhongguo Jingji Nianjian 1982 (China Economic
Yearbook), Beijing 1983).

Production of large oxygen-generating equipment came to a halt,
because no new orders were received. In 1980 and 1981, only 3 and 1
sets, respectively, of sizes above 1,000 Nm3/h were produced
(Zhongguo Jixie Gongyedi Fazhan (Development of China's Machinery
Industry), p. 91). At the time, apart from Hangyang, China had
several factories producing cryogenic equipment, the largest of
which were in Kaifeng and Harbin (both capable of producing 3,350
Nm3/h oxygen equipment) and in Sichuan. With capacity standing idle,
production at the Hangzhou plant was diverted to the manufacture of
beer bottle sealers, round frames (for packaging), 70 mm. gas pipes
and other consumer-oriented goods. At the same time, Hangyang
entered into a license agreement with Hitachi for the manufacture of
medium-pressure oxygen turbo-compressors for the 10,000 Nm3/h
models, which resulted in successful trial production three years
later.

Readjustment and reorganization by the government

As part of the readjustment plans, the Ministry decided to create a planning and controlling organization, the China National Air Separation Equipment Corporation in 1982. Its personnel, almost one hundred engineers and technicians and administrative staff, came from eight factories producing cryogenic equipment. However, most of them had split off from the Hangzhou factory. Until the 1984 reforms, this company had been invested with controlling powers over the entire branch of industry in China. It tried to coordinate the Chinese market and make sure that each of the eight factories had enough work. With capacities standing idle and factories running at a loss, cooperation in planning and consultancy was thought to be beneficial, especially to the smaller enterprises. The CNASEC was financially supported by the Ministry of Machine-building Industry. Early in 1984, its membership was expanded by the addition of the Shenyang Blower Factory, the Lanzhou Vacuum Equipment Factory, the Fourth Electric Equipment and Machinery Installation Company (of the Ministry of Machine-building Industry) and the Design Institute of Jilin Chemical Industry Corporation. Thus it strengthened its capability to organize the production of complete plants.

In November 1984, however, as part of the economic reforms which were meant to increase the decision-making powers of enterprises, the CNASEC lost most of its administrative powers and became a commercial company. Since then, it has become a service company which supplies organizational and technical services of designing, engineering, and assembly, and it contracts for the supply of air separation and other industrial gas equipment and appendages by Chinese or foreign producers. On behalf of the Ministry, it also builds and manages the Suzhou Welding Technique Training Centre, due to open in spring 1989. The company now has some 130 personnel, started paying taxes in 1987 and says it now operates without government subsidy. It seems that in the past few years, the company's human and organizational resources have been underutilized.

In addition to the creation and subsequent abolition of the CNASEC as a controlling organization, the government decided to transfer ownership of Hangyang to Hangzhou Municipality early in 1985. This further reduced the links between the CNASEC (still based in Hangzhou) and Hangyang. According to the management of Hangyang, this did not affect Hangyang's ability to obtain foreign exchange or negotiate with foreign companies.

Further measures: import restriction, licensing and a working division.

In the same year, the government took three more measures regarding the industry.

First, imports of oxygen-generating equipment of sizes of 6,000 Nm3/h oxygen were restricted. The last such import had been purchased, from Linde, by the Hunan Provincial Economic Committee on behalf of the Xiangtan Steel Plant. The government felt that Hangyang had sufficient capacity and quality to offer. Thus, imports of these sizes had to obtain prior approval from the Ministry of Machine-building.

Second, a license to produce pressure containers of the second category was given to three more factories in 1984, and in 1986 all eight factories, except Wuxian, were licensed to produce first second and third category pressure containers. In 1984, the license for the third category (issued by the Ministry of Person-nel and Labour, which was charged with safety inspection) had been given to the Hangzhou factory only. Thus, Hangyang lost its temporary monopoly position.

Third, the government imposed a new working division between the various plants in the industry. Its main outline was as follows:
-- Hangyang was planned to become the producer of large-scale units, based on cooperation with Linde. Linde was now approached with a demand for joint production of 30,000 Nm3/h units. The Hangyang investment programme under the 7th Five-Year Plan (1986-1990) was directed at such production. Hangyang was forced to transfer its

knowledge and documents of the 6,000 Nm3/h units it had been producing for many years to the Harbin Oxygen-generating Equipment
Factory.

-- the Harbin Oxygen-generating Equipment Factory was a producer of
a wide range of small size (mainly 60 Nm3/h N and 50Nm3/h O2) air
separators, centrifugal ventilators, air compressors, pumps and
granulators for the sugar industry and some other products. By 1985,
it had 2,000 employees, of which only 160 were technicians and
engineers - very few, considering its range of products. With RMB 27
million of fixed assets, its turnover was RMB 17 million (Zhongguo
Dongbei Jingji (Economy of Northeast China), Beijing 1987, vol.III,
p.197).

-- the Handan Oxygen Generating Machinery Factory in Hebei, with
2,500 employees, was specialized in the production of 7 models, with
capacities of 50, 100 and 300 Nm3/h oxygen (with purities of 99.5 -
99.6 %) and 100 Nm3/h high-purity nitrogen (Zhongguo Gongshang Qiye
Minglu 1982-1983 (Directory of Chinese Industrial and Commercial
Enterprises), Beijing 1983, p. H 12-13).

-- The Sichuan Air Separation Equipment Factory, in Jianyang, with
2,500 employees, manufactured several small-sized oxygen, nitrogen
and argon producing units, and 1,000 Nm3/h, 3,350 Nm3/h and 6,000
Nm3/h air separators. Production of the latter was, it seems, halted
by the Ministry. It was a main producer of various sizes of
cryogenic liquid containers and started a new production process of
150-liter containers in 1984, on the basis of a license from the
Japanese Daido Oxygen Company. It also designed and manufactured a
new line of LPG equipment and started a new small (150 Nm3/h) air
separation unit, capable of argon extraction, in addition to its
existing 50 Nm3/h units. Furthermore, it continued producing small
cryogenic compressors, pumps and valves.

-- The Jiangxi Air Separation Machinery Factory in Jiujiang, with
900 employees, produced a.o. 150 Nm3/h air separators, with a high
quality, energy saving, medium-pressure turbine expansion machine.
Plans were prepared to further develop its 10 and 20 ton
refrigerated containers and its 3 ton liquid trucks.

-- The Zigong First Machinery Factory in Sichuan, with 1,400
employees, was a producer of small air compressors, liquefactors,
LPG bottles and small (6 - 120 Nm3/h) air separators. Together with

the China Air Separation Equipment Company, it developed and trial-produced 20-inch cold storage containers.

-- The Kaifeng Air Separation Equipment Factory, with 3,000 employees, remained directly under the Ministry. It had a similar wide range of products to Hangyang, producing air compressors, small oxygen compressors, low-temperature pumps, liquid containers and valves. In 1984, it designed and manufactured, for the first time, a 2,000 Nm3/h high-purity nitrogen generating unit and a large-size LPG tank. In 1985-6 it imported cold storage techniques and a production line for large-sized (300 to 1,000 tons) cryogenic containers from West Germany (EMB Company). It developed a highly efficient double-axial centrifugal air compressor. Moreover, also with government support, it developed its moulding technology for polyurethane foam. Although it produced only medium size air separators, of between 1,000 and 6,000 Nm3/h, by 1986 the factory claimed to have a leading position in China with its high-purity nitrogen installations with molecular sieves. It then sought to improve its design and manufacture of plate-fin heat exchangers through foreign cooperation (Liu Guogang, Lin Zongdang (ed.), Zhongguo Jingji Jishu Xiezuo Shouce (Handbook of China's Economic and Technical Cooperation), Beijing 1987, Vol. I (5) p. 329. On behalf of Kaifeng, the Ministry opened negotiations with the French company Air Liquide and the American company Air Products, in order to discuss joint production of large-size equipment of sizes between 6,000 and 30,000 Nm3/h at Kaifeng.

How serious these negotiations were at the time is difficult to say, but they still continue. Air Products has been given to under-stand that they can only sell in China if they enter into some kind of cooperative arrangement with the Kaifeng plant. However, there are serious doubts about the capacity of Kaifeng to develop production of such large-size equipment. Certainly, cooperation might push the sales of its appendages, such as its air compressor for 10,000 Nm3/h units which has been found acceptable by Hitachi. Also, the negotiations may be meant to put pressure on Linde and Japanese companies. The negotiations with Linde and the other companies on joint production were still going on in December 1988, and we will revert to them later.

The technology transfer between Hangyang and the Harbin Company

The imposed technology transfer from Hangyang to Harbin met with the greatest reluctance from Hangyang. Cooperation was kept down to a minimum. No training was given, no explanations were offered of the documents, and requests for support were ignored. There were, of course, good reasons for such an attitude. For one thing, Hangyang felt it would be helping the Ministry to create another competitor for large-size equipment. This reflected a certain lack of confidence in its future position as a producer of large-scale equipment of sizes of 10,000 Nm3/h. The company seemed to doubt ots own qualities and/or the size of this market. The efforts of the Ministry to turn Kaifeng into a producer of the same sized equipment in cooperation with other foreign companies worried the Hangyang management still further. Secondly, the Ministry had assigned the task of technical consultancy and guidance of Harbin in the manufacture of this new product to the CNASEC, which seemed fair enough: the CNASEC had been created by the Ministry precisely for this purpose and was packed with engineers and technicians drawn from Hangyang who had all the necessary experience with the production of the 6,000 Nm3/h types. Thirdly, no payment had been given for the transferred documents. Thus, the uncooperative attitude, which continues until today, was based on self-interest and lack of enthusiasm about the way the Ministry had handled the industry. At the end of 1988, the Hangyang engineers were no longer worried about Harbin; they believed Harbin would not be able to move to the production stage on the basis of the documents alone, even with the help of CNASEC, for many years to come.

The government measures mentioned above must be seen against the background of a renewed vigorous growth of the steel industry and the oxygen-producing equipment industry since 1983. In 1981-1982, the government had intervened to reduce the effect of the crisis; in 1984-1985, it intervened to plan the industry's future growth. Neither of the interventions appeared to be very successful.

The state of the industry in the mid-1980s

Around 1985, there were still 8 specialized factories which produced gas separation and liquefaction equipment. In addition to the plants mentioned above, there were the Handan and Wuxian Oxygen producing equipment factories (both produce small-size oxygen generating equipment), the CNASEC and two specialized research institutes for air separation and liquefaction equipment in Hangzhou and Sichuan. All plants but Wuxian had permits to design and produce first, second and third types of pressure containers. Each had a somewhat different range of products. The Hangzhou factory, which had expanded from a little over 4,000 personnel in 1978 to 5,700 personnel by 1985, was the largest by far, and the most complete, because it also produced large air and medium-size oxygen compressors, low temperature pumps and valves and other appendages. In the mid-1980s, employment in the industry increased rapidly, from 14,600 employees in 1984 to 21,100 employees in 1986; of whom, the number of engineers and technicians had increased even more rapidly, from 1,100 in 1984 to 1,900 in 1986. Output and turn-over had increased even more. Value of gross production (expressed in 1980 prices) more than doubled, from RMB 99 million in 1984 and RMB 123 million in 1985 to RMB 215 million in 1986. Sales income climbed from RMB 149 million to RMB 246 million between 1984 and 1986. Profit and taxes of the industry amounted to RMB 36 million and RMB 16 million respectively in 1986, up by one third over 1985. For more financial details see Table 2. First, however, we will consider

Table 1. OUTPUT OF CRYOGENIC EQUIPMENT IN CHINA, 1984-1986

(tons)

Type of equipment:	1984	1985	1986
Air separation and liquefaction	7,387	8,199	10,568
Multi-component air separators, liquefactors	431	113	162
Cryogenic liquid storage and transport	2,836	3,778	3,233
Acetylene dissolution	95	284	690
Other branch products	909	2,367	2,716
TOTAL	11,658	14,741	17,419

Source: Zhongguo Jixie Gongye Nianjian 1987 (China Machinery
        Industry Yearbook 1987), p. III 24-25.

output.

The very crude indicator of weight alone shows a 49 percent output increase between 1984 and 1986 (see Table 1). If we compare this with the increase in output value (in 1980 prices) of 117 percent, we may conclude that there was at the same time a considerable shift to higher-value. This applies both to the air separators and to the fast growing sector of "other branch products", which includes air and oxygen compressors made by Hangzhou and some other plants within this branch. None of the above figures include production of appendages (compressors, valves, filters, pumps, meters etc.) by other branches of the Ministry of Machine-building Industry, or by enterprises under other ministries.

The speed of development also showed in investment figures. The original value of the industry's fixed assets was RMB 281 million at the end of 1986. In that year, RMB 103 million had been invested in capital construction and technical transformation. The economic indicators began to show improvement. Its performance was in line with most of the machine building industry, a branch which still lagged behind the economic performance of China's industry in general. The productivity of the machine industry in China has been stated to be 10 to 20 times lower than in the USA or Western Europe (Sun Xiaoliang and Zhang Jianmin, "An analysis of the economic performance of machinery industry", Zhongguo Gongye Jingji Yanjiu (Research on China's Industrial Economy) 1988:5). Table 2 gives some data on the air separation equipment industry.

Table 2. ECONOMIC PERFORMANCE AND ITS INDICATORS IN THE CHINESE AIR SEPARATION EQUIPMENT INDUSTRY, 1986

Gross output value :	RMB 243 million (current prices)
	RMB 215 million (1980 prices)
Sales income :	RMB 246 million
Net output value :	RMB 82 million
Profits :	RMB 36 million
Taxes on sales :	RMB 16 million
Profit cum taxes handed over to the government : RMB 18 million	

```
Fixed assets value :              RMB 281 million (original value)
                                  RMB 147 million (after depreciation)
Stipulated floating capital: RMB 185 million
Output value per RMB 100 of fixed assets        :   RMB 81
Profits per same                                :   RMB 18
Capital : profit-cum-taxes ratio                :   11.6 percent
Output value per labourer                       :   RMB 10,198
Average circulation period of floating capital  :   275 days
```

Source: <u>Zhongguo Jixie Gongye Nianjian 1987</u>, p. III 210-211.

In spite of the improvements achieved in the mid-1980s, actual production lagged considerably behind the industry's potential. The Hangzhou plant co-produced and produced only 6 units of the Linde-type 10,000 Nm3/h type between 1978 and 1988, the first four of which were almost completely made by Linde. Furthermore, from 1982 till the end of 1986 it produced 11 sets of the older 6,000 Nm3/h type (<u>Hangzhou Nianjian 1987</u> (Hangzhou Yearbook 1987), p. 85).

Renewed Hangyang demands for cooperation with Linde

In the early 1980s, Hangyangs design and production methods had remained virtually the same, with the exception of an improvement of the controls, now mostly DPC's (digital processing controls). In 1984, Hangyang started using molecular sieves (which had not been in the contract with Linde). In the same year, Hangyang demanded that the 13 months of training at Linde left over from the 1978 contract, be used for training of their engineers in the newest technological developments with respect to processing design and molecular sieves; in this way, it hoped to be able to upgrade the model it had acquired in 1978. It claimed that under the contract, Linde should have supplied them with the newest techniques. The chemical factories in China, which in the 1980s became a more important customer group than the steel industry, were only satisfied with the newest technology. Linde, however, pointed out that the Chinese had not even fulfilled their existing obligations under the 1978 contract of jointly producing and selling eight complete plants.

After the first four plants contracted for in 1978-1979, the fifth and sixth plants had been contracted for and produced by Hangzhou in 1983-1985 without any items manufactured by Linde. Linde did supply some special machinery and instruments, however. Hangyang paid its due royalties. Since around 1982, China was able to manufacture the air and nitrogen compressors needed for these plants. Yet the oxygen compressors still had to be imported, and the clients often preferred foreign appendages anyway and were prepared to pay for them. The vague wording of the contract had produced ill feeling on both sides. After the 1981 crisis, "eight sets of equipment in the period 1979-1982" had been interpreted by the Chinese in a rather far-fetched manner as meaning that no obligations existed from 1983 onward, irrespective of the number of sets produced. Although subsequently Hangyang modified this position, as indicated by its payment of royalties, the damage had been done. In 1985, Linde sold a US$ 3.3 million 10,000 Nm3/h oxygen component package (with rare gases) for the Benxi Steel Company to CNASEC through CNITC, and the remaining components were contracted directly by the Chinese buyer with Hangyang.

In 1985 and 1986, Hangyang delegations visited Linde in Germany again, asking for future cooperative production, and mutual understanding improved. In October 1985, Hangyang publicly announced that Hangyang and Linde had entered into an agreement for long-term cooperation in the production of 30,000 Nm3/h units during the 7th Five-year Plan, and this was written in Hangyang's subsequent sales brochures, too. But cooperation remained limited to the fulfillment of one order for a 30,000 Nm3/h oxygen producing unit by Shougang (the Capital Iron and Steel Factory in Beijing). For this unit, Hangyang delivered a minor part of the equipment (30 percent of total weight, but 6 percent of total value), notably the cooling towers, the molecular station and the shell of the cold box. We will refer to this contract below. Although hailed by the Chinese as the beginning of a new era of cooperation in production of this modern large-size equipment, up till the end of 1988 no new similar orders have been obtained by either Linde or Hangyang.

The 10-year cooperation agreement between Linde and Hangyang expired at the end of 1988. However, the agreement had a clause which stipulated an automatic extension. The Hangyang management was anxious for a renewal of their cooperation. It believed that only through such cooperation would the two companies be able to withstand the growing competition. However, the form of cooperation would be different from the old. Linde was asked to invest in the setting up of a new cooperation. The 30,000 Nm3/h models to be produced here would not only serve the Chinese needs, but could also be sold on the international market. Because of the low labour costs, production costs would be lower than in Germany. The Chinese suggested that, if Linde had any doubts about their product quality, overseas sales from the new factory might be limited to Third World countries. Such worries about quality were unfounded, they believed, because the Chinese aluminum welders were quite good. Linde's investment would mainly consist of training and documentation for the new models. Particularly needed from Linde were the software and methodology for the necessary design calculations for high-pressure oxygen plants with molecular sieves (of over 85 Bar, instead of the regular 30 Bar) demanded for chemical factories.

If these proposals seemed too ambitious to Linde, Hangyang indicated that it was quite capable of producing a larger part of the 30,000 Nm3/h units than it had under the Shougang contract. Hangyang could manufacture the heat exchanger, more appendages, and also part of the interior of the rectification column (in fact, Linde had already tested, and approved for use, one such part made by Hangyang).

Discussions about a new period of cooperation and technology transfer between Linde and Hangyang were still inconclusive at the end of 1988. Unlike the previous contract, Hangyang now asked Linde to provide its newest technology and its largest equipment, and to provide venture capital for a new plant. Thus, in the short term Linde would obtain no financial benefits from its transfer of technology, but would have to put up capital instead. Judging by the past performance of Hangyang, and considering the market situation and the difficulties of joint ventures in China, this request can hardly have been attractive to the German company. However, Hangyang was convinced that the two companies were condemned to work together, as neither was able to beat foreign competition for the large-size models single-handed. The Linde management had not expressed its ideas about future cooperation very clearly. Possibly, it would prefer such cooperation to take the form of Hangyang being a subsupplier in Linde sales contracts, for those items which either the customers or the Chinese government demanded it should supply. In such a case, without a definite order from a Chinese customer, cooperation could not take a concrete form.

Competition from various foreign companies

Before 1986, Linde and the Japanese companies of Nippon Sanso (Oxygen), Kobe and Hitachi had been the only competitors on the Chinese market, but since then, Air Products and also the Soviet Union has joined them. In reaction to the increased competition and the anticipated expansion of the Chinese market, Linde established a small office in Beijing in 1986.

Around 1985, Kobe Company had supplied and installed two sets of 14,000 Nm3/h oxygen generating units in the Baoshan steel plant. However, actual performance of the two sets fell short of the required amount of 28,000 Nm3/h by 3,000 Nm3/h. So, the Baoshan steel factory demanded compensation. Finally, after mediation by the Hangyang management, the Japanese supplied and installed an extra unit free of charge. For Baoshan, this meant a delay and somewhat higher operating costs (including a higher electricity bill). The Japanese name, which had not been too good before, was hurt by this. But then, their prices, which often were some 10 per-cent lower than Linde's, already reflected the quality difference. Because of increased competition and the soft-loans introduced by the Japanese since then, profit margins have gone down for all companies.

Starting from 1986, the Soviet Union sold three 14,000 Nm3/h oxygen generating units to the steel factories of Handan, Tangshan and Taiyuan, respectively, on the basis of barter trade. According to the Hangyang management, Hangyang could have produced these 14,000 Nm3/h units just as well - it would have produced better quality than the Japanese and more modern sets than the somewhat outdated Soviet types. Hangyang reacted by offering its large-size installations for sale to the Soviets and to Hungary, but so far without success.

The Anshan and Baoshan steel factories both purchased a 30,000 Nm3/h oxygen installation (priced at over US$ 20 million) from the American Air Products Company (which was obliged to cooperate with Kaifeng) in 1986-7. The Wuhan Steel Factory was still negotiating with Linde, Air Products and Air Liquide in late 1988. According to Hangyang, Air Products would be hard to beat: with the low dollar, this company offered prices which were below those of Linde and Hangyang, both for the main units and for additional equipment. Air Products had taken some 20 Chinese Americans into its service, who started promoting sales in a more active manner than had been usual in the industry in China before. For example, in connection with the negotiations, a group of Wuhan managers was invited to come and visit the United States for free.

The problems of Hangyang

The Hangzhou plant still has not achieved the same reputation for reliability with Chinese customers as Linde has. When the sixth 10,000 Nm3/h unit produced by Hangyang did not perform well, the end-user contacted not Hangyang, but Linde, about what he presumed to be a mistake in the original Linde design. The Linde engineer discovered a minor flaw in the steel welding of the air purification unit. The leaks were repaired subsequently by the Hangyang people. As for the 6,000 Nm3/h units, since the early 1970s the mainstay of the Hangyang sales programme, in spite of their lower prices, Chinese customers have often preferred foreign-made ones as being more reliable. Below, we will refer to these consumer preferences in three specific cases.

More seriously, the Hangyang production of the 10,000 Nm3/h models in the 1980s lagged far behind original expectations. The lack of orders was due to several external factors: the 1981-1982 crisis of the steel industry, the subsequent demand from the chemical industry for larger and improved models (with molecular sieves), and an investment programme in the Chinese steel industry that seemed to favour either larger installations (such as in Baoshan, Beijing and Wuhan) or smaller ones (of 6,000 Nm3/h). The government efforts at reshuffling the air separation equipment industry must have caused some customers to shy away from the Hangzhou plant. On the other hand, the import restrictions instituted in 1985 provided Hangyang and the other factories with virtual monopoly positions in their respective market segments for equipment of 10,000 Nm3/h and below. The lack of orders was also due, however, to various internal factors at the Hangzhou plant, which caused it to lose its competitive edge.

For one thing, its range of products and varieties was and is very large. Many of these took large investments of capital, skilled manpower and time to develop, but have only a very limited market in China. This goes particularly for its large-size air and nitrogen compressors, the medium-size oxygen compressors, the molecular sieves, and low-temperature valves. The plant has some 1,200 machine tools for cutting, casting, welding, electric plating et cetera, and, like so many Chinese machinery factories, prides itself on being capable of producing almost anything. This has made it difficult to give the necessary attention to the improvement in production methods and design of the 10,000 Nm3/h units.

When questioned on this point, however, the Hangyang management stated that making money was not their main goal. Products were and still are priced on a cost-plus basis. It expressed its belief, however, that economic results would not improve by a limitation of their product range. Instead, lowering of costs should be achieved by an improvement of production methods. By 1988, 80 percent of the aluminum welding was still done by hand; if a new agreement with Linde for the joint production of 30,000 Nm3/h units were to be concluded, it might become possible to acquire automatic welding machinery. The management wants to extend, not to contract, its range of products, to all appendages of the large air separation units with the exception of magnetic valves. Notably they would like to produce large oxygen compressors (for which they are the only Chinese manufacturer) for the future 30,000 Nm3/h units, on the basis of a cooperation agreement with the European companies DMAG or Sulzer. One may doubt, however, how wise such an effort would be. Most likely, the foreign manufacturer would license only one type and size of compressor, the Chinese market for which would be very small indeed.

Second, delivery times took longer than scheduled. Foreign contractors can usually order their raw materials, equipment, parts and accessories from various independent producers and suppliers, and obtain what they need quickly. In contrast, Hangyang depends for its raw materials (aluminum, steel etc.) and main products on supplies from about 30 to 40 different Chinese companies, and some foreign ones, over which it exerts very little control. Although contacts may be direct, actual supply goes through semi-governmental channels. In the case of foreign imports, permits are required. With other Chinese companies, approval by the relevant authorities takes some time. Worst of all, there is very little possibility of taking legal action in the event of late delivery or of changing to another supplier. This imperfect internal market organization goes a long way towards explaining why the Hangzhou factory wishes to extend the scope of its own products and become less dependent on subsuppliers.

Third, for the major raw material, aluminum, a severe shortage has existed in China for a long time. Since 1984, the share of aluminium obtained through direct allocation by the central government has dropped to less than 40 percent (usually, this allocation was linked with the designated key-point projects of the Five-year Plan), and to make up the remainder, Hangyang had to find and purchase aluminum on the "free" internal Chinese market (which is dominated by government agencies). The shortage and unreliability of the electricity supply was another external limiting factor. On average, it reduced daytime production by 30 percent, and the Hangzhou plant has had to rearrange its production shifts to two per day. It did not receive permission to generate its own electricity, probably due to the fact that the government's development plans for Hangzhou stress its future as a recreational area, which should be protected from further industrial pollution.

Fourth, as with many other factories in China, the Hangyang management does not aim at profit maximization, and cost awareness at Hangyang is still underdeveloped. Rather, its primary goals are technical progress, extension of the scope of supply, expansion of production and the achievement of a greater reliability and predictability. With the imperfect operation of the Chinese market, achievement of the latter goal is sought through a reduced dependence on outside suppliers. It is clear, though, that the factory did not and does not generate enough profits to be able to invest in the development of its many products. Above, we noted that the factory did not even invest enough to be able to make full use of its one big foreign purchase: the 10,000 Nm3/h oxygen-generating installation. Possibly, the shortage of funds may have caused delays in production, which in turn made the company less attractive to potential customers.

Financial considerations are held to be of a lesser importance. Apparently, the management has not made calculations of relative profitability, for the simple reason that this is not used as a basis for decision-making. Production planning and marketing are underdeveloped. The RMB 20 million investment programme for 1986-1990, agreement about which was reached with the authorities at the end of 1985, was based on certain assumptions about, but not conditional on, future profits. Once accepted, the Hangyang management may have felt it had little to worry about. The cost-plus price setting method would insure that planned profits could be achieved.

Following cost increases due to inflation, higher wages and prices of materials, between 1985 and 1989, the sales prices for oxygen-generating equipment have doubled. At the end of 1988, Hangyang charged RMB 15 million to RMB 17 million for its 6,000 Nm3/h unit, up from a price of RMB 6.5 million in 1984. Still, the Hangyang plant claimed to be some 20 percent cheaper than its foreign competitors. The steep rise in the price of raw materials since 1986 increased the value of the aluminum stocks held by Hangyang. In 1987, the official list price rose only modestly, from RMB 6300 to RMB 6800 per ton. But increasingly, aluminimum was traded at much higher prices than the list price. The financial results of Hangyang in the past few years look satisfactory on paper: an annual increase of 20 percent. However, the increase is mainly due to inflation. See Table 3.

Table 3. GROSS OUTPUT VALUE, PROFITS AND TAXES OF
HANGYANG, 1986-1989 (million RMB)

	Gross output value	Profit cum taxes
1986	60+	12
1987	72	14
1988	82	16
1989 (planned)	100	20

These figures may have contributed to an optimistic expansive mood on the part of the management and government authorities. Such a mood was further strengthened by the satisfactory order position of the plant, which by the end of 1988 had guaranteed work for 18 months. However, its financial improvement was based mainly on factors other than economic performance, and the self-generated funds for investment remained behind original expectations when corrected for inflation. The originally allocated amount of RMB 20 million of investment over the period 1986-1990 was a small sum, a mere 5 percent or so of the total expected turnover. Because of inflation, it has become even more inadequate. Thus, the expansionist Hangyang plans lack a sound financial basis.

STEEL INDUSTRIES: PLANNING, NEGOTIATIONS AND PURCHASES OF
EQUIPMENT

The Chinese steel industry and its requirements for oxygen-generating equipment

The Industrial Survey of 1985 showed that China had 71 medium and
large size iron and steel factories, with a total of 1.6 million
employees and RMB 30 billion of fixed assets (without deprecia-tion,
RMB 45 billion). During the 6th Five-year Plan (1981-1985), these
enterprises had invested RMB 22 billion, most of which in the
purchase of equipment. About two thirds of this sum had been
allocated by the government (the Ministry of Metallurgical Industry)
from profits and taxes handed over to it, but a growing percentage
of this was paid by the enterprises out of retained profits.

Table 4. FINANCIAL RESULTS OF LARGE AND MEDIUM SIZE IRON
AND STEEL ENTERPRISES IN CHINA, 1980-1985
(billion RMB, 1980 prices)

Income from sales	21.0	27.4	33.8
of which, tax on sales	1.3	2.3	3.9
factory costs of sold products	15.5	19.4	24.0
of which, depreciation	1.0	1.4	1.8
profits	4.4	5.7	5.7
of which, handed over (partly as tax)	3.8	3.6	2.6
of which, retained by enterprises	.6	1.6	2.3

Source: Zhonghua Renmin Gongheguo 1985-nian Gongye Pucha Ziliao,
(Materials of the PRC 1985 Industrial Survey) Vol. 1, Beijing 1987.

In 1986, only 12 percent of the investments in the steel industry
were financed through state budgetary allocations. 50 percent was
financed by the enterprises themselves, and 25 percent through loans
(Dong Yizheng, "Handle the five relations in the development of
steel industry correctly", Zhongguo Gongye Jingji Yanjiu (Research
on China's Industrial Economy) 1988:3).

The steel industry had already invested in the purchase of native and foreign equipment on a very large scale during the 1970s, and continued to do so in the 1980s. According to the Industrial Survey of 1985, for steel making alone (i.e. exclusive of iron smelting, rolling and composites) the 71 large and medium sized companies had 2.2 million items of equipment, one third of which had been imported. Of these 762,000 items, 314,000 had been imported in the 1970s and 267,000 (or 13 percent) in the 1980s. The percentage of foreign imports is higher than with most other Chinese industries. The chemical industry and coal industry showed a similar share of foreign-imported equipment, but much less of it was imported during the 1980s. In the chemical and machinery industries, the imports of the 1980s represented only 5 and 4 percent, respectively, of its total equipment in 1985. Steel has been a priority item with the Chinese government over the past decades, and this was reflected in its foreign import programme.

However, there is a considerable lead time between investment and production, often of five years or more. On the basis of investments realized in the early 1970s, between 1975 and 1980 China's steel output rose from 24 million tons to 37 million tons, but then output stagnated till 1983. However, the demand for steel increased in line with the unexpectedly rapid growth of the economy. Imports of steel rose dramatically from 5 million tons in 1980 to 20 million tons in 1985, and in money terms, from US$ 2.2 billion to US$ 5.8 billion. The slowing down of investments during the so-called Readjustment Period of 1979-1981 now began to cost the government very dear. Although steel output rose from 47 million tons in 1985 to 56 million tons in 1987, it could not keep up with demand. For the period 1986-1990, it was expected that the annual shortage of rolled steel was to increase from 18 million tons to 28 million tons, all of which would have to be imported (Wang and Gong, "A preliminary study of supply and demand of steel products under the 7th Five-Year Plan", Jingji Yanjiu (Economic Research) 1987:1). Mainly because of the energy shortage and transportation problems, steel output even went down in early 1989. Thus, during all of the 1980s, steel was in short supply and the steel companies operated in a sellers' market. Although most of their sales still have to go through the state plans, with prices fixed by the government, a small but rapidly growing percentage has been permitted to be sold directly at higher prices since 1982 and since 1984 more of the profits may be retained. The case of Shoudu Iron and Steel Company (Shougang) has been put forward as an illustration of the positive effects of this so-called "contract system" (Zhou Guanwu, "On the contract system", Jingji Guanli (Economic Management) 1986:3). Since then, prices and profit margins have gone up. In 1987, the official price for common steel rose from RMB 670 to RMB 840 per ton, and its market price from RMB 1,400 to over RMB 2,000 per ton (Shandong Jingji (The Economy of Shandong) 1989:1)

In line with the decentralization policies, and also because of the problems encountered with Baoshan and other large projects undertaken by the central government, in the 1986-1990 period China has emphasized a simultaneous and independent development of all of its medium and large scale iron and steel plants. Just as before, however, all plans for expansion require approval from the State Planning Commission and the Ministry. There was (and nowadays still is) a general feeling, that steel plants should not become too large, in order to avoid problems with a long construction period, supply of iron ore, long-distance transport of coal, and finance. Soviet advisors have suggested that for steel production in China, plants with a capacity of one million tons would be most appropriate. In the early 1980s, most plants had an annual production capacity of between only 200,000 and 700,000 tons of iron, often with a somewhat lower capacity for steel making. Their productivity was low, and has not improved much in recent years. In 1985, one half of China's steel was produced by top-blow convertors, and the other half still by open-hearth furnaces and electric ovens. At the end of the 1970s, most factories had received approval to modernize and expand their steel production by 200,000 tons or more by 1985. A dozen national large-size steel plants, mostly with a capacity of 1 to 3 million tons, were scheduled to increase their production capacity by between half to one million tons. On that basis, the former plants demanded oxygen-generating installations of 5,000 to 6,000 Nm3/h, while the largest plants needed larger capacities of between 15,000 and 30,000 Nm3/h. Because of readjustment policies, in 1980 most of these plans were shelved, to be taken up again in 1982. Under the decentralization policies, the necessary investment had to be paid by each enterprise itself, from its retained profits. If foreign exchange was needed, this had to be included in the national or provincial plan, because the steel industry could claim that its products substituted imports, thus saving foreign exchange. Because of the decentralization, implementation of the previous plans for expansion now came to depend mainly on the financial position of the enterprise, the drive of its management and the support of the Provincial Economic Committees. Once expansion with a definite capacity of steel making had been decided upon, the additional need for auxiliary facilities such as oxygen-generating equipment automatically followed. Only rarely would a plant decide to replace existing equipment (the official write-off period for oxygen-generating equipment is 18 years) or create reserve capacity for future expansion.

For the foreign suppliers, but also for Hangzhou, China's decentralization meant a total change in their sales. In the 1970s, they had been selling batches of equipment to the national import corporation in Beijing, after negotiations with the Ministry. The 1978 contract for four and eventually eight sets under the Linde-Hangzhou agreement was still an example of this centralized approach. In the early 1970s, for reasons of national security, or just because of a general secretiveness, foreign companies were not even told which plants in China would be the end-user. For instance, in 1972 Kobe Company sold nine sets to China, which were allocated by the Ministry to various steel plants throughout China. The Lianyuan Steel Plant in Hunan received two of these 1,500 Nm3/h sets, although it would have preferred another type. The Japanese were not allowed to go to Lianyuan municipality, which was not open to foreigners at the time, and a joint inspection of the boxes with equipment took place in Changsha, the capital of Hunan. That was the first and last time Lianyuan had dealings with Kobe. The design had been made by engineers from the plant itself, and the equipment was installed by a company designated by the provincial branch of the Ministy (the Non-ferro Third Construction Company in Zhuzhou (Hunan)). There could be no overseeing or subsequent inspection by

Kobe. Instead, Lianyuan invited inspectors from a plant in Xinjiang, where a similar set had performed very well. The Japanese paid RMB 290,000 in compensation when an air compressor appeared to have a rusted axle. They paid the same compensation for the same flaw to Jinan (Shandong) without demanding inspection, although there the boxes had been opened in the factory without any Japanese being present. However, there were no problems at all with performance later. Such barriers between supplier and end-user were unsatisfactory to both sides, and have been partly, but not totally, removed since 1978. As a consequence, sales contracts became individualized, for one set only. As noted above, in the 1980s some companies made a beginning with active marketing.

The suppliers of oxygen-generating equipment do not play a role in planning. On the basis of an approved expansion plan, the steel plant fixes its specifications of capacities and intended scope of supply. It collects environmental data (climate, wind direction, quality of air and water) and establishes electricity supply and voltage. Usually, prior to approval of the plan, location and design have already been decided upon, with or without the help of a specialized design institute. It then invites would-be suppliers to bid. During contract negotiations, however, some minor amendments may be made by the customer on the basis of the information received. The supplier then proposes his design at a preliminary design meeting.

Since 1978, overseeing by the (foreign) supplier of installation work and training have become standard parts of the contract. Construction is always done by a Chinese company. Large steel plants often use their own construction company.

Selection of oxygen-generating equipment: procedures and Criteria
Over the past ten years, managers of China's steel plants have become more independent and their technical expertise has greatly increased. Decisions about imports of technology, however, which involve a great deal more than technical aspects alone, have gone through successive waves of decentralization (1978 and 1984) and centralization (1980 and 1985). Steel plants under different

ownership, of different status or size (in descending order, "national keypoint", national, "local backbone", local large-size, local medium-size) found themselves in a different position vis-à-vis the State Planning Commission and the Ministry of Metallurgical Industry (both used their own import organization, not MOFERT, for imports of whole sets of equipment), the Provincial Economic Committee and other local authorities. In the following, we present three cases of imports by steel plants of different status or size:

- the Capital Iron and Steel Company (Shougang) in Beijing is a large keypoint enterprise, and one of the most advanced and profitable in China. Between 1980 and 1985, its steel output increased from 1.5 to 2.6 million tons.

- the Xinagtan Iron and Steel Company (Xianggang) in Hunan is much smaller (about one fifth of the size of Shougang) and more dependent on outside support. Technical performance indicators put the plant at the bottom of the list of the national keypoint enterprises, and its economic results have not been too good, either.

- the Lianyuan Steel Company (Liangang) in Lianyuan, also in Hunan, is a local backbone enterprise of a somewhat smaller size than Xiangtan, technologically inferior but with very good economic results.

All three plants imported oxygen-generating equipment in the 1970s and 1980s. Together, their experience may show how Chinese companies introduced foreign technology under different constraints. First, we will sketch the usual sequence of sales of oxygen generating equipment in China from the perspective of the foreign company. This brief sketch applies to keypoint projects approved by the government. Local projects not included in the national plan, particularly if the seller is asked to accept a buy-back arrangement, are much more complicated. In the last few years, half a dozen steel companies have been allowed to import directly using their own funds (Shougang is one of them), but all others must apply to the government for permission.

In the 1970s, negotiations used to fall into two parts. First, there were technical discussions about the scope of supply and technical data, leading to the signing of technical annexes by the Chinese design institute (on behalf of the Ministry), contracting agency and sometimes end-user. This was followed by commercial negotiations with the contracting agency. In the 1980s, the two negotiations were intertwined, and the main contacts became those with the end-user.

First, the foreign company is approached unofficially by the end-user. It is asked to supply technical data, for the benefit of a feasibility study which the end-user is preparing. Sometimes, the end-user unofficially asks for technical advice, too. After the steel company has completed its technical feasibility study, suppliers of equipment may be asked for a preliminary technical proposal. Subsequently, three or four suppliers are usually asked to give a presentation and make a preliminary bid. This may be followed by an inspection visit to the bidders´ factory, by a delegation composed of representatives of the end-user, design institute and commercial organization. Food and lodging is usually provided by the foreign company, air travel by the end-user (the same is true when such a delegation visits a Chinese supplier). This is followed by a technical clarification meeting. Then, the bidders make their technical and commercial proposals (since 1986, these have had to conform to a standard form set by China National Technical Import Corporation TECHIMPORT).

Finally, formal negotiations start, with all bidders at the same time but in different rooms. The technical annexes are discussed, and the foreign companies try to reach a signed agreement on technical matters, as quickly as possible. The commercial negotiations may run parallel or start a little later, and culminate in a final round of "financial gambling" (as one participant put it). Conditions of payment are fairly standard: FOB, 15 percent downpayment, balance upon delivery, penalties for delays and under-performance, acceptance testing and joint signing of a completion certificate, a one year guarantee on performance and materials, and a final payment of 3 percent. There may be a commercial loan or a soft loan. The contract, once signed, has to receive formal government approval, a process which takes from between months to one year.

The suppliers´ proposed price may be considerably reduced in these final rounds of negotiations. In order to meet the Chinese demands for a lower price, the foreign companies are usually forced to reduce the items ´supervision´, ´training´ and ´spare parts´. The Chinese customer cannot change his minimum requirements of quantities and purities of oxygen and nitrogen, but he may lower his demands for high-quality appendages, delete foreign items for which there are Chinese substitutes, or reduce the scope of supply in other ways. However, the Chinese side prefers to reduce the service element and the spare parts first. It is a common experience that the supervision period has to be extended later. Both sides also know, that shortening the training period will cause problems, although the Chinese negotiators may be rather optimistic about the learning abilities of their engineers and operators. However, the Chinese attitude is not to touch problems now which will emerge and be solved later. The enterprise strives to fix a low price for the initial investment in the contract, and cares less about later running costs. Later, during performance testing, the Chinese engineers are very concerned about purities and quantities of oxygen and nitrogen, but do not usually question energy consumption. This emphasis on the initial investment during negotiations reflects the financial situation of Chinese enterprises: it is difficult to get approval and funds for investments (especially foreign exchange), but there is no shortage of circulating capital within the enterprise. Moreover, the Chinese negotiators are usually squeezed between a low project budget and high technical demands. Both, the investments approving agencies and the engineers play major roles in the negotiations. Thus, concern about later costs to the enterprise is pushed to the background.

During negotiations, the two parties are rather different in size and quality. The foreign company may have two or three representatives, with back-up in case special problems arise. Of course, they know their own equipment and what it can do in great detail. In most cases, they are capable of going into any proposal by the customer for technical adaptations or changes. The Chinese customer, in contrast, has a negotiating team of 15 to 20 people. "Team" may not always be the appropriate word, as the members represent the different interests of several departments, they may raise minor departmental points, and as a group may quarrel amongst themselves during negotiations. The technical participants come from the steel enterprises´ departments of planning, instrumentation, electricity, and from the operation group of its oxygen plant. Their qualifications are unequal, and few, if any, grasp the entire process. Thus, they need a great deal of explanation. According to one participant, "technical negotiations often become training sessions". Of course, the language barrier is a problem, somewhat less so since the mid-1980s. The interpreters often lack the necessary technical understanding. However, the Chinese participants come well prepared. They often have already visited similar installations in China, and solicited opinions there. During negotiations, they frequently consult their ducumentation, which contains standards set by the government and translations of the technical specifications of the foreign equipment, made available to them by the Ministry. Thus, the end-user can and does draw on a nationwide network of information. In the experience of Linde, the Chinese negotiators show great interest in process details, but less in equipment design. Yet, during the discussions, the opinion of the design institute is often most decisive. The foreign company prefers not to prolong the technical discussions, mainly because its engineers are expensive and in demand elsewhere. On the Chinese side, the speed of the negotiations depends largely on whether a deadline has to be met. The Chinese stand to gain from an elaborate comparison of all options, the expert explanations they receive improves their technical knowledge, and they also generally like to show their hospitality and devote time to their guests. The Chinese staff often has ample time, and can "play a home match". Under these circumstances, the adage "time is money" applies only to the foreign company.

Shougang

Shougang already had five sets of oxygen-generating equipment in operation, both foreign and Chinese-made, when it needed an additional large-size installation with a capacity of 30,000 Nm3/h of oxygen for its steel expansion program in 1986. The five were

-    one 6,000 Nm3/h Kobe set bought in 1960 and installed

in 1964, which was completely Japanese
- two 6,500 Nm3/h Air Liquide sets installed in 1966 and 1967, with some Chinese appendages
- one 6,000 Nm3/h Hangyang set installed in 1976, with five Shanghai-made air compressors
- one 3,350 Nm3/g Hangyang set from 1959, which was due to be taken out of use.

The company had been doing very well financially. (Actually, it had so many more funds available than it was permitted to invest in steel and steel products, that it started to buy up other enterprises). Thus, it could afford to buy the best. The company wanted to produce high-purity oxygen, in order to further improve its already high standards of steel making, but it also wanted argon and rare gases. It wanted quality performance of the current international level. At an early stage, it asked a Linde engineer to look into their needs and help to formulate requirements. A senior Shougang engineer, when visiting Europe for other programmes, also visited Linde, Air Liquide and an Italian steel factory. The two European companies and Air Porducts were then asked for proposals. The preliminary design was commented upon by the Shougang's own design department (the company felt it did not need people from the Beijing Central Institute of Design and Research for Iron and Steel), and all demands could be met.

Of its selection criteria, Shougang considered four to be the most important:

1) the technical level. This included several aspects: the purity of the gases produced, their quantities, and energy consumption. Linde scored best on the first two points, but not on energy. For the plant, energy is an important cost factor (in 1987, in addition to 110,000 kWh per day produced by the company itself, it had to purchase about 90,000 kWh, 20,000 kWh of which against a high price);

2) the scope of supply. The company wanted the best quality of electrical equipment. It had been very satisfied with the Siemens motors in the Air Liquide sets, and with Siemens service in China. It also wanted high quality oxygen compressors, and preferred Sulzer. For the latter, Linde had proposed a German vendor as a

cheaper alternative. The Ministry of Machine-building Industry had insisted that Hangyang should supply a substantial part, and had contacted Linde about cooperation. This would reduce the forex component, which was something wanted by the important corporation. Hangyang would receive a processing fee directly from Shougang, without having to pay foreign exchange (hereafter forex) to Linde. Shougang complied with this government policy;

3) the price. Shougang did not start from a fixed budget, but did want to reduce costs, especially the foreign exchange. The government insisted that Shougang should demand a loan arrangement for the forex component;

4) the delivery date. Shougang wanted the installation to be delivered and installed as soon as possible.

On the basis of the above criteria, the Shougang negotiators seemed to have already made up their minds before the final round of negotiations with the three companies started. Linde could meet and even surpass the technical demands, had promised to supply the Siemens equipment, and had included cheaper oxygen compressors in its design. A deal had already been worked out by the Ministry, which guaranteed local (Hangyang) participation (manufacture of cooling towers, part of the molecular sieve, shell of the cold box and some other parts, totalling 30 percent of the weight and 6 percent of the contract price). Because Linde and the Shougang design institute had designed according to the state norms, there was no need to go through the inspection authorities' approval procedures, and Linde could supply within a year. For the production of rare gases, Linde enjoyed great confidence. Thus all that remained to be settled was a reasonable price and a loan arrangement. The price had been lowered through a reduction of site services; the 80 months of assistance in China (notably for installation) originally proposed by Linde were reduced to 40. For training in Germany, Linde had not made a suggestion of its own, but it accepted the proposed seven months for training in welding, operations and calculations. It was the Ministry, not Shougang, which insisted on a loan. Linde could not promise the type of soft loans which might be given by its competitors with support of the Japanese, French or British governments, but for its first time in

China agreed to give a commercial loan with a reduced interest rate. This would defer payments by three to seven years. Thus the contract was concluded.

In order to be able to evaluate the investment decisions of the Xinagtan and Lianyuan steel companies, we first have to look at their rather difficult position in Hunan Province. Hunan is an inland province with a shortage of both high grade iron ore and coke. Two thirds of its iron ore must be imported, from Dabaoshan (Guangdong Province), Hainan Island and Australia. It is transported mainly by rail over about 1,000 kilometers or via the Yangzi River (Xiangtan can be reached by 500 ton boats in winter). Local ores contain too much phosphorus, zinc and other minerals. In the 1980s, about three million tons of coke was imported every year, from Pingdingshan and the provinces of Henan and Guizhou. The province has a severe shortage of electricity, too, particularly in the dry season (55 percent of its electric power supply is from hydroelectric stations) (Hunansheng Tongjiju, Hunan Gongye Jingji Ziliao Huibian 1949-1985 (Collection of Materials on the Industrial Economy of Hunan), Changsha 1986). Unlike the costal provinces, its export earning are very limited, so it has little foreign exchange. Its technical level is below other areas, too. Thus the development of Hunan`s steel industry was and is severely limited by its location, raw materials supply, and energy. Xinaggang, for that reason, receives political support from Beijing; this was illustrated by the visit of Li Peng, then minister of Metallurgy and Energy, in 1987.

Yiangtan and Lianyuan steel plants together produced over 90 percent of Hunan`s iron and steel. Table 5 summarizes production and financial data for both companies and Shougang in 1980 and 1985.

Table 5. PRODUCTION AND FINANCIAL DATA FOR XIANGGANG,
          LIANGANG AND SHOUGANG, 1980 AND 1985

		Xianggang		Liangang		Shougang
		1980	1985	1980	1985	1985
Iron output	(1,000 t.)	?	528	300	400	3,259
Steel output	(,,)	?	555	300	396	2,579
No. of employees	(1,000 p.)	21.8	23.1	15.3	16.5	123
Turnover (1980 prices, million RMB)		362	337	153	241	1,861
(current prices, ,,)			441		354	n.d.
Net output value (current pr. ,,)		105	160	54	138	1,229
Profit + tax (,, ,,)		75	59+59	40	70+44	814+287
Value of fixed assets (after depreciation) (million RMB)		318	376	192	229	1,580
Liquid assets (,,)		9.2	31.4	45.6	79.2	483
Productive investments 1981-5 (,,)		– 98 –		– 38 –		n.d.
Investments for renewal and transformation (,,)		23		34		230
Energy use (coal eq., 1,000 t.)		1,360	1,025+)	679	689+)	n.d.

+) 1984 figures

In comparison, the local Liangang company showed a much better
performance than the national Shougang company. The latter had a 7
percent decrease in turnover between 1980 and 1985, if measured in
constant prices. The former had an increase of 55 percent, while its
physical output of iron and steel increased by one third. The local
company had a higher profit ratio, and could dispose of more liquid
assets. The difference was partly due to the very limited product
range of Xianggang, which in accordance with the national state plan
had to supply mainly steel bars and rods fore wires, at State-fixed
prices. In contrast, Liangang supplied very diverse products to meet
the local needs of Hunan. Also, Xianggang was more dependent on
national allocations of raw materials. Yet productive investments of
both were very limited during this period. Only after the end of
1984, once the decentralization policies had created more
independent local decision-making and steel prices had started to
rise, was a vigorous investment program implemented at both plants.
At that time, the main capacities of the two plants were as follows:

Xianggang had two modern 750 cu.m. high furnaces, with an iron melting capacity of 600,000 tons; three open-hearth furnaces for continuous steel making, also with a capacity of 600,000 tons; a primary rolling mill and a rolling mill for steel wire materials, with a capacity of 450,000 tons; two 45-hole coking ovens.

Liangang had its own mining, producing 200,000 tons of iron ore per year; a sintering and pelletizing capacity of 380,000 tons; two 42-hole coking ovens producing 560,000 tons of cokes annually; two 255 cu.m. high furnaces and one 84 cu.m. high furnace, with a total iron melting capacity of 400,000 tons; two 15-tons top blow convertors (a third was added in 1986), three 6-ton and two 1-ton electrical ovens, with a total steel making capacity of 400,000 tons; six rolling mills, with a 350,000 ton capacity for steel materials.

Xianggang had already obtained approval from the Ministry of Metallurgical Industry for an expansion of its steel production to 750,000 tons in 1990 in the 1970s; later, a goal of one million tons was approved for 1993. Three open-hearth furnaces had been installed in 1980. Each could produce only 180,000 tons of steel per year without additional oxygen. With liquid oxygen, the factory expected that each furnace would be able to produce 400,000 tons, or even up to 500,000 tons, by raising the utilization rate and reducing the refining period. There would be considerable savings, of 30,000 tons of heavy oil per furnace annually.

Because of the crisis, the investment program was interrupted, but in 1983, Xianggang, together with the Wuhan Steel Designing Institute, compiled a plan book for the equipment of one furnace with 6,000 Nm3/h oxygen-generating facilities, which laid down all technical demands in detail. Yianggang sent the plan to the Ministry, together with a cost estimate, early in 1984. For the cost estimate, the factory had based its calculations on the price of similar (French) equipment installed in Maànshan in 1976 – a mistake which would be regretted later. The Ministry quickly approved the requested size, quality and the need for import, but it did not provide the foreign exchange. At that time, decentralization measures had just been taken which shifted responsibility for concluding foreign contracts from national agencies such as TECHIMPORT to provincial governments (or rather to their import corporations), up to a financial limit. For the first year of 1984, Hunan Province had been allocated a quota of US $ 100,000,000 for imports of large-size capital equipment. The Leading Group of foreign Negotiations of the Hunan Provincial Government decided to include the Xinagggang proposal in the half a dozen projects which it submitted for formal approval to the State Economic Commission in Beijing. A representative of this commission then came to the provincial capital of Changsha, discussed the projects and gave his final approval. The entire approval process was carried out quickly, and Xianggang was able to invite bidders in mid-1984. Never before, and never thereafter, would it be so easy for Xianggang to obtain foreign exchange.

Xianggang invited three Japanese and three Western companies to come to Changsha and make proposals. Of these, according to the Xianggang management, Hitachi did not come because it felt competition was too intense, and Air Products (which had been approached in Beijing) did not consider the project to be of sufficient importance. Hangyang was never asked for a bid, because the quality and reliability of its 6,000 Nm3/h models were considered to be inferior. At an early stage of the technical discussions, Air Liquide was feared to be rather expensive, because of its use of copper materials, and the Kobe equipment was felt to be outdated technology. Thus, Nippon Sanso (Oxygen) and Linde were the chief competitors.

At the beginning of the technical negotiations, Xianggang`s main selection criteria of were:
1) reliability. This would be the first oxygen factory, and there was no back-up. The management wanted a two-year guarantee period, and equipment which would not give any problems;
2) advanced techniques and high quality. Because this would be a foreign import, the chief mechanical engineer demanded advanced international technology. He wanted high purity oxygen and nitrogen, in order to improve their usually rather low quality of steel. He wanted a molecular sieve plant, and two expansion boosters, which

would be more economical in energy use. He wanted automated
controls, not manual ones. The valves should be electromagnetic and
computer-controlled;

3) the highest quality appendages were demanded: Siemens electrical
equipment, DMAG air compressors and Sulzer oxygen compressors. All
these demands had been discussed within the factory in advance,
agreed upon and been included in the project document;

4) a 100 cu.m. liquid tank should be included, and 150 bar bottling
equipment with over-the-counter sales facilities.

However, it immediately appeared that the formulated demands were
impossible to meet within the available budget. Instead of
reconsidering, or applying for additional foreign exchange, which
most likely would have meant a postponement of a year or so, the
Xianggang negotiators chose to trim the project while maintaining
its high-quality technical core. Among other things, they agreed to
cut out argon production facilities, they reduced the spare parts by
almost one half, to seven percent of the contract value, they
adopted Chinese-made cooling water equipment, perlite, nitrogen
compressors and some meters, and accepted a substitution of cheaper
AEG electrical equipment and GHH air compressors.

When commercial negotiations started, with three of the four
companies, Xianggang had already revised its requirements of the
basis of the technical negotiations. Linde had the advantage of
having three highly qualified engineers on the spot, who could
suggest a number of cheaper alternatives to the Xianggang people.
The Sanso negotiators did not have the same flexibility, and had to
consult their company (and be absent) rather often. Xianggang liked
Linde's suggestion of increasing the oxygen generating capacity to
6,200 Nm3/h; its equipment had a higher oxygen extraction ratio than
Sanso, of 92 percent as against 88 percent. Linde had come up with a
better method of gas compression for the filling of oxygen bottles,
and suggested cheaper appendages. Now that the available budget of
foreign exchange would have to be fully used anyway, the Chinese
technical people also dominated the commercial negotiations. The
Chinese had set themselves a deadline: the Chinese New Year's Day of
1985. Thus, negotiations were concluded with uncharacteristically

fast speed within three weeks. According to the chief mechanical engineer, three weeks of negotiations were normal for such a project in their company, which does not like drawn-out negotiations. Possibly, there were budgetary reasons as well: the foreign exchange was part of the 1984 allocation to Hunan Province. The Chinese negotiating team decided to opt for Linde. The Chief Mechanist, who had led the negotiations, reported their results orally to the Factory Board, and got immediate approval without questioning. Thus the decision was taken. Four years later, the company was still satisfied with its choice, though it recognized that the trimming down and substitution of cheaper appendages had caused some trouble.

Liangang had been founded during the Great Leap Forward in 1958, but was dismantled after two years. In 1965-6 it had been reopened because of the "Third Front" strategy of locating vital industries in inland mountain areas. Liangang lies some 100 miles west of Changsha in Loudi Municipality. At the end of the 1970s, the State Economic Commission and the Hunan Provinvial Planning Committee had approved its expansion plans for a steel output of 400,000 tons in 1985. In 1980, Liangang made a concrete investment plan, but this was shelved. In 1982-3, the Hunan Metals Planning and Desin Institute did the planing and design for the necessary additional oxygen-generating equipment. A joint report was submitted to the Provincial Economic Committee, for approval of size (5,000 Nm3/h), importation, and maximum amount of foreign exchange, and at the same time to the Changsha Ferro Mining Design Research Institute for approval of design. Both approved. Now Liangang itself could select the equipment.

Liangang`s main considerations were:
1) the price. In the years 1983 till 1985, Liangang had a very substantial investment program for expansion of its iron and steel production. The funds had to be provided by the provincial government, and been budgeted in 1983, when steel prices were still low. Therefore, the budget was meagre. It had been agreed that Liangang would pay back the foreign exchange it received from the provincial government in Renminbi, at a rate one quarter higher than the official exchange rate;

2) reliability and service. Liangang was very satisfied about the performance of the two 1,500 Nm3/h Kobe sets which it had purchased in 1972. The Japanese had provided an excellent repair service since then. However, in 1984, the Ministry had warned Liangang that the Kobe sets in Baoshan did not work well, and recommended buying Sanso equipment (Sanso had sold a 3,500 Nm3/h set to Ankang, which performed quite well, and had also contracted for 5,000 and 6,000 Nm3/h sets in Liuzhou, Lanzhou and Laiwu Steel Plants);

3) the original quality demand for the oxygen was a purity of 99.3 percent, which was equal to the state minimum norm for steel making, (Yejin Gongyebu (ed.) Gangtie Oiye Zhiliang Guanli (Quality Management in Iron and Steel Enterprises), Beijing 1984, pp. 305-6). (For welding and forging, this purity suffices only for second-grade steel; for first grade steel, the norm is purity above 99.5 percent). If the installation could produce oxygen with a higher degree of purity, so much the better. The lower quality demand for oxygen would also save electricity. However, all foreign companies guaranteed higher purities, from 99.6 percent upward. The factory did not have high demands for the nitrogen, because its steel did not have to be top quality. Because it used local ores, the phosphorus content of its steel was very high (0.5 percent) and the silicon content (0.7 percent) was high, too (Gangtie (Iron and Steel), 1987:6). In order to reduce investment costs, no liquid oxygen tank had been planned (the existing oxygen-producing sets would serve as a partial backup, if the new set`s oxygen supply were to be interrupted). However, the company had demanded a liquid argon tank and gasification unit. With a lower liquid oxygen requirement, the expansion turbines did not have to be highly efficient. The lower quality demands also made it possible to use manually operated controls, valves etc.;

4) the selection of a simple technology would reduce the need for service and training. Liangang had sent 16 people to Beijing for training in the 1970s, and received two specialized graduates from the Beijing Steel Institute, and one meter specialist from Northeast University. These had been adequate for the operation of the existing oxygen-producing sets, and would be for the new sets.

Why didn't Liangang consider a native supplier, such as Hangzhou? Hangyang at that time needed such orders badly. For one thing, the demanded size of 5,000 Nm3/h was larger than the 3,350 Nm3/h models and smaller than the 6,000 Nm3/h model currently produced. The company and, apparently, the design institute and the Ministry doubted the ability of Hangyang to adjust its own designed 6,000 Nm3/h model. The company, moreover, may have expected better service and greater reliability from a foreign company. At least, this had been shown in a comparison between its Kobe sets and a same size set supplied by the Kaifeng Plant in 1971.

In July 1984, at the Hunan Provincial Foreign Contract Meeting, Liangang met with representatives of Kobe and Sanso. Subsequently, discussions about a contract were started with them, as well as with Linde and Hitachi, in Changsha. But Linde and Hitachi never had much of a chance. The standard Hitachi sets were too small (3,200 Nm3/h) and Liangang didn't want to install more than one set (it might have bought Chinese-made sets in that case). The Linde proposal was much too expensive: there was a difference of US $ 1.5 million between their offer and the final contract price (820 million yen plus 98 million yen for spare parts, CF) with Sanso. Kobe and Sanso gave detailed information about the quality and capacity of their products, but the other two companies did not get so deeply involved. The quoted energy consumption of the Sanso set was low (0.56 kWh per cubic meter of oxygen). Kobe was well known, and Sanso had been recommended.

The final choice fell on Sanso, mainly because of Liangang's worries about the performance of Kobe. The contract specified the types and brands of air and oxygen compressors (Toshiba TIKE 250, and J4D 3003 with synchronous motor), but for most appendages only type and performance were mentioned. The contract did not provide for any training, but it included four months of overseeing of installation by four Japanese specialists. There was a twelve-month guarantee period. As with Xiangtan, installation work was contracted out to a Huanese company. The price had been low, but Liangang was going to have many problems.

CONTRACT EXECUTION AND SOME LESSONS TO BE DRAWN FROM EXPERIENCE

Execution of the contracts: training, supply, overseeing installation work, trials and performance testing.

Of the three companies, Shougang had the installation work done by
its own construction company, but for the rare gases it contracted
an installation company under the Ministry of Aviation. The two
Hunanese steel factories contracted provincial construction companies, which had been designated by the provincial branch of the
Ministry. The foreign companies provided the basic design data for
construction. The construction workers did not need training, but
installation workers did. Foundation work, with slabs of concrete,
was done well, and in the case of Xiangtan, where an extra
insulation layer was added for the mounted tanks, even extremely
well. The latter construction and installation was held up by Linde
as an example to other Chinese customers - a source of pride for the
Xiangtan management. All projects were basically turnkey: the steel
structures for the equipment, the ducts and pipes, the wiring
materials etc. were pre-fab delivered by, and the responsibility of,
the foreign company (although parts of the first were Chinese-made),
together with drawings and installation sheets. The construction
companies operated under standard government contract rules, which
stipulate prices for labour and materials, and also entitle the
customer to compensation in case of faults or delays.

There was a big difference in the quality of installation work and
its supervision by the foreign company in the three factories. Also,
experience with training and performance of the equipment supplied
varied a great deal. Liangang was least satisfied, and rightly so.

In Shougang, performance tests had gone well and the management did
not report any particular problems. There were few faults with the
equipment and appendages supplied by Linde and Hangyang, and these
were remedied without delay or other difficulties. The purity of
oxygen, as noted on the daily operation sheets, appeared to be 99.9
percent (at a variation in hourly supply between 1,200 and 6,700
cu.m. of oxygen). However, for the rather complicated rare gases

equipment, the German engineers had cause for some dissatisfaction with the installation company and with the level of the trainees, and Shougang should have paid more attention to communications. The Germans started enthusiastically, but did not get enough response. As the overseer explained:

"When I arrived, I decided to give all the Chinese one month of extra training from 5 to 7 in the afternoon, because they were not capable enough. This was not in our contract at all... We should have had some colleagues of a similar level as us, or some good technicians to work with, but there hasn't been anybody. In Taiyuan and Wuhan, where we worked before, the situation was much better. People were more friendly and collegial. Here, we have no counterparts with whom we can talk. Their is no point in being an instructor this way. I told this to the senior Shougang engineer who is in charge, right at the beginning, but got no reaction. Now we just bide our time..."

In spite of how they felt, the two were quite active in giving training classes, to almost thirty people in different shifts, and even gave English classes. Compared with trainees in other third world countries, however, the Chinese had achieved less. Both blamed this on their educational background and work environment:

"Chinese trainees are afraid of making mistakes or admitting to them. Therefore, they seem opinionated. They seldom dare to adjust the gases and try to experiment, because they are afraid something might go wrong (which would affect purity or quantity of the oxygen supply). Because there are too many personnel for one job, they do not concentrate. Three hours of concentration is their maximum.

His German colleague was introduced to the responsible Shougang engineer only after three months. The limited professional and social contacts, which may have been partly due to the fact that the installation work was done by an outside company, caused them to worry about aspects of their work. For instance, the Chinese had refused to say from which local factory they would purchase their test gases, once the Linde supply ran out. The German engineers were anxious about their quality, and offered to have local samples analyzed, as this is of the utmost importance for proper analysis by the gas chromatographs. The Chinese obviously felt it was none of their business and that local test gases would be good enough. They knew the future supplier would be the nearby Beijing Municipal Oxygen Factory, but didn't tell the Germans. The installation company people (a group of six to eight technicians) had left without even asking the German overseer whether their job had been completed, although it had been stipulated in the contract that he should sign a completion certificate. Possibly, the installation company had reasoned that their obligations were to Shougang, and not to the Linde engineer. In sum, as far as rare gases installation was concerned, Shougang had neglected communications with the foreign experts, and not made maximum use of their presence. For the main installation, no problems were reported.

In Xiangtan, construction and installation work had been quite satisfactory. The cold box had been mounted in a extraordinarily level position, and insulation was very good. Xianggang had trained a dozen aluminum/iron welders for four months, and all their subsequent weldings ("there were over 800 connections") were first class. Five of these welders would later be sent to Shougang for assistance. Some unforeseen problems had been successfully overcome. The overseas training of eight man/months of personnel, which had been part of the contract with Linde, had to be cancelled after the government forbade such foreign trips in 1985. Instead, Xianggang sent installation and operating personnel for training to the new oxygen-manufacturing installations at the Jinshan Petroleum Company in Shanghai and the Ma'anshan steel works. Some computing engineers were sent to a university in Beijing, where they were trained in the use of a computer similar to that used by Linde. This saved money for both sides, without seriously affecting results.

Only the trial tests took longer than expected, which may or may not have been due to lack of previous instruction at the Linde plant. At least according to Xianggang, this was due to the attitude of the three German engineers, who did not give clear instructions or listen carefully to Chinese opinions. After the second trial had failed,

"we held a meeting and said: you must write down your methods, so that we can discuss them. We must cooperate better! We should divide our duties. You must lead us, but we must understand the details!"

With adapted operation procedures, the third trial was successful. A minor problem arose, when a safety inspector of the Ministry of Labour found slacks on the welding surface of a steel tank, and did not pass it. Approval was received after rewelding by Xiangtan. Furthermore, because of their being in short supply or because of long customs procedures (the Xianggang management did not know which) the Siemens electric motors arrived somewhat late. Installation was completed in February 1987, and actual production started in April, both on time.

There was still some trouble with equipment, however, and Xianggang held back some of the contractual final 3 percent of the payments after the guarantee period had ended. Because of the high temperatures in Xiangtan, of over 40 degrees Celsius in summer, the motor of an air compressor burnt out. Linde supplied another one (of the same brand), but it still did not operate satisfactorily at high temperatures because of cooling problems. Also, the Austria-made motor of the refrigerating units had already burnt, and been replaced, three times. According to Xianggang, this would not have happened if they had bought Siemens motors. But then, it had been their own proposal during contract negotiations to substitute these and other appendages for cheaper ones. The large liquid oxygen pump shook too much, and installation of crossbars did not solve the problem. It had not worried the foreign engineers before, but Xianggang wanted it to be looked into again.

The performance of the installation, both in quantity and purity of oxygen, was quite good. However, on inspection the control sheets showed that heavy and prolonged drawing in the downstream factory causes the purity of oxygen to drop to levels of 98.3 to 98.7 percent. There is pressure on China's steel factories to produce as much steel as possible, to relieve the national shortage, and the price differential for high-quality steel is too small. "This hampers technical progress, renewal and replacement of products, and affects the positivism of the enterprise's production management" (Yejin Jingji Yanjiu (Research in the Economy of Metallurgy), Vol. 19).

Summing up, the Xianggang engineers were satisfied with their cooperation with Linde, very satisfied with its design, and reasonably satisfied with the equipment, with the exception of the refrigerators.

In Liangang, the contract had not provided for training by the Japanese Sanso Company. In the years 1972 till 1978, Liangang had 16 employees trained in Beijing, and recently three specialized graduates from higher educational institutions had been assigned to Liangang. The Zhuzhou Installation Company did its work very quickly, and sloppily. Two years later, the shell of the cold box was a sorry sight; it was leaking and even had holes in several places. Some ducts leaked, too. Overseeing of installation work was also different from Xianggang. At Xianggang, there had been one German overseer, who stayed for six months, with additional specialists being flown in now and then. At Liangang, overseeing lasted only two months, and was finished 52 days earlier than stipulated in the contract. Sanso sent four overseers for welding, gas and electrical work, and brought in two specialists for the air compressors and two for trial and performance testing. The latter four had not been mentioned in the contract, and Liangang did not have to pay for them. Because of the speedy completion, both sides saved money. However, the trials were unsatisfactory. According to Liangang, the two Japanese did not know what to do, and had to continuously call Tokyo head office. "Then we operated the machinery ourselves, and within 12 hours we solved the problems of air separation. Our people did better than the Japanese!" The contract stated that quality checks should take one month, but actually they took three. However, in spite of the bad results the attitude of the Japanese overseers was much appreciated, even later! They took ample time to deliberate with the Chinese and always asked their opinions.

Every day, there was a joint meeting from 5 to 6 p.m. to discuss results of the day's work.

Liangang was confronted with serious problems, some of which have not yet been solved. The major one is that production of oxygen has to be limited, in order to achieve a satisfactory degree of purity of oxygen, and the purity of nitrogen is also below standard. The management could not make explain whether this was due to original design failures, or to faulty installation or equipment. During my visit in November 1988, it appeared that nitrogen purity was not recorded: the nitrogen meters had broken down six months before. They had not yet been replaced, because this fell under the Sanso guarantee, and Sanso had been slow in delivering. The Chinese were afraid that Sanso would not pay, if they were to replace them on their own initiative. With the argon, and the cooling water and gas temperature, performance was good. In 1985, several of the delivered appendages appeared to be defective, when the boxes had been opened for joint inspection, or developed trouble and needed repairs soon afterwards. The Japanese just lowered the price twice, by 32 million yen. Liangang had to accept, because it knew that replacement of the equipment would take a very long time. The equipment would have to be changed in Japan, and go through all customs procedures again. The expansion turbine had to be repaired by the Sanso people within a year, the oxygen compressor once before and once after the one-year guarantee period had expired, the nitrogen meters could not be repaired and were removed, but not replaced (see above). For all those reasons, production of oxygen suffered a great deal. The management was quite dissatisfied, and contrasted its newest purchase with its smaller Kobe units, which had been running so well for such a long time.

The various steel factories had several common features:
-- They had selected foreign oxygen-generating equipment without seriously considering Hangyang, but after collecting many opinions from other steel factories, the design institute and the Ministry. Preparatory planning and obtaining of approval had taken a considerable time, and all wanted to decide quickly about which equipment to purchase and did so.

-- There was no specialized installation company, and all had to train aluminum welders and other installation personnel, most of whom came from their own personnel. Often, they had been prepared for their training to a certain extent, by active or passive participation in similar installation projects at other steel plants. However, the foreign overseers usually feel that there are too many trainees, and that personnel is changed too often. With each new project, they have to start training all over again. Supervision becomes more difficult, when the Chinese workers are not well educated, but believe they are. On the other hand, the work schedule is always reasonable. There is little financial incentive to produce good quality work, so one must rely on eagerness to learn and professional pride. However, after the job has been done, the professional skill acquired is often lost because of lack of practice.

-- During installation and trial, as a rule technicians from several other steel factories were present, to learn or to assist - usually without the foreign supplier and overseers being aware of the presence of these outsiders. In this way, very valuable experience was exchanged, with beneficial results. One wonders why the foreign companies did not use these opportunities for training and local assistance more. Possibly, their presence in China was too limited, in time and in number of people, to organize such cooperation.

-- Personnel on the payroll of the oxygen plants numbers at least five times as much as would be considered normal in the West: Shougang has about 400, Xianggang 140, Liangang 215. Yet in the control rooms, the usual number of people is 3 or 4, working in three shifts, which is in accordance with foreign recommendations. There are separate maintenance crews (e.g. 70 people at Shougang). Most of the staff is obviously working elsewhere or absent. This situation of overstaffing is common in the steel industry. As one author stated: "at the present level of management and organization of our steel factories, combined enterprises which produce one million tons of steel annually need 20,000 to 30,000 employees...In old factories, this is not very different from new ones" (Yejin Bao (Metallurgical Bulletin), July 16, 1985).

-- Running costs of operating personnel, and also energy consumption, were never considered of much importance. Possibly, the iron and steel industry is already such a large consumer of electricity (e.g. in Hunan it consumes 11 percent of the total electricity output) that oxygen generation hardly counts.

-- "What you pay is what you get" may not always be true, but Xianggang later regretted that it had been forced to buy cheaper quality appendages than originally planned and to trim down its installation, and Liangangs experience with buying cheaply is most unhappy. In both cases, the original budget was set too low. This can be attributed in part to insufficient foreign contacts and foreign experience, a legacy of the past, which China was overcoming slowly during the 1980s. However, inflexibility in the Chinese system itself is also to blame. The companies had no authority to change the forex budget previously approved by the provincial government. This has changed since 1987: Shougang and Xiangtan can now import directly and attract foreign loans themselves.

-- In their planning of investments and implementation of projects, all three enterprises have suffered from the government's stop-go policies regarding investment approval and foreign imports. With rising steel prices, investments in the steel industry are quickly recouped. In Xianggang, the costs of the new oxygen-generating set had been recovered after three and a half years. However, the changing government policies have introduced an element of uncertainty to dealings with foreign companies, be it training, acquisition or investment.

As we have demonstrated above, both for internal and external reasons the Hangzhou Oxygen-generating Equipment Manufactory (Hangyang) has not made full use of its ten-year cooperation agreement with Linde A.G. Production and sales of the 10,000 Nm3/h model have remained much below original expectations. Over the 1978-1988 period, there were only minor improvements on the model. There has not been sufficient progress in independent design, computation and production techniques. The main factors responsible for this seem to be the following:

-- Hangyang bought a license for one model, but did not make an effort to develop from there. Ten years later, in the 1988 negotiations, it again asked for a license for the production of one model (of 30,000 Nm3/h capacity), this time without giving any consideration to the training and development of the know-how of its own engineering and designing staff. In the early 1980s, the interruption of sales within China because of the crisis led to a severance of professional contacts with Linde. There was little business to be done, and the Chinese may have felt ashamed to confront the German engineers with their troubles. Yet in this way the Hangyang engineers slowly lost the benefits of the cooperation and of their training in Germany. The lack of a proper computer exacerbated their problems in developing or even maintaining their capacity for independent designing. Certainly, if Hangyang wants to become competitive on the world market (and in China!), it should be able to make its own calculations about performance and to redesign equipment within the cold box.

-- China has too many factories, all of them medium-sized, which produce similar or identical oxygen-generating equipment. This makes for high production costs. Although most companies may not care too much, because they can calculate their prices on a cost-plus basis, it makes it very difficult for each individual enterprise to concentrate sufficient human and financial resources on product development and large-scale production for a larger market. This goes not only for oxygen-generating equipment, but also for gas exchange valves, air compressors, oxygen compressors, liquid tanks etc. With dozens of medium and small sized factories, it is difficult to raise quality, increase output, lower costs, and beat foreign competition.

-- Most medium-sized factories, such as Hangyang, manufacture too wide a range of products. Each factory makes its own standard types, for many years or even decades in succession. Political and economic pressures on factory management to cater primarily to the various needs of the provincial, or even municipal, market may be too strong to resist; the same authorities control the supply of many vital raw materials and machinery. However, it gives the technical staff of the factory insufficient time and funds for the development of superior products or production methods. The commercial staff in most heavy industries has been accustomed to being in a sellers' market for a long time, and this has reduced their motivation to search for and develop new markets for high quality products. The financial staff is burdened by cumbersome procedures for obtaining investment funds, loans or foreign exchange. The personnel department is under pressure to hire as many personnel as the factory's budget can stand.

Taken all together, the Hangyang management has had to operate under severe administrative limitations. Product diversification was a logical response, which contributed to their need for greater independence from unreliable outside suppliers and allocations through administrative channels. With the growth of a genuine market economy in the past few years, it would make more sense to concentrate on large scale production of a limited range of products, where Hangyang could be a market leader. The experience of the past decade should be summed up, and lessons be drawn from it. In the future, technology transfer might involve a greater use of foreign expertise and foreign cooperation, and self-sustained development of quality products for local and foreign markets.

Beijing, Hangzhou, Changsha,
Xiangtan, Lianyuan, and Leiden,
November 1988 - January 1989.

Part Three

China and the World
in the Nineties

A SUMMARY OF GLOBAL TECHNOLOGY TRENDS OF
POSSIBLE STRATEGIC INTEREST TO
THE PEOPLE'S REPUBLIC OF CHINA

Prof. Lewis M. Branscomb
John F. Kennedy School of Government
Harvard University
Cambridge Massachusetts 02138, U.S.A.

Introduction

This brief paper will first summarize some important trends in science and engineering that are influencing the innovation process and diffusion rates of technology, the requirements for rapid progress in science, and likely effects on technical cooperation and competition.

In the second section a number of specific developments or trends are selected that may be of interest to specific industrial sectors.

The third section enumerates some strategic choices based on the changes overtaking the processes for technology generation and the diffusion process.

Significant trends

1) The impact of science on engineering

For many years engineering was considered an art. It drew from science; but intuition, experience, and testing out ideas in working models introduced an important empirical character to the innovation process. With the maturing of science - first physical and now biological science - and with the availability of computers for simulation, engineering has become much more predictive and quantifiable.

Europe-Asia-Pacific Studies in Economy and Technology
Leuenberger (Ed.) From Technology Transfer
to Technology Management in China
© Springer-Verlag Berlin Heidelberg 1990

The ability to describe an innovation in a computer model, specifying all the relevant material and process properties, means that many designs can be evaluated without recourse to prototyping or "bread board models". That descriptive capability, in turn, follows from great advances in scientific instrumentation and materials characterization.

All elements of the innovation cycle - design, development, production process development, production, and testing - depend directly on current information in science, and are increasingly interdependent. The linear model of the innovation process (basic research yields ideas which are explored through applied research and engineering development, and then transferred to the factory for production) is no longer an accurate description. Important consequences of this merging of science and engineering are:

a) a reduction in the time required for a new product idea to reach production,

b) designs and processes that take full advantage of scientific possibilities, with narrower design margins but higher performance,

c) the necessity of organizing research, development and production in more closely linked ways,

d) the necessity of a merging of scientific and engineerring "cultures", with impacts on scientific education,

e) with electronic communication, a facilitation of the diffusion of designs and processes when these are quantitatively documented in computers,

f) with communications and computer controlled production tools, the rapid diffusion of manufacturing processes to similarly equipped plants,

g) a greater risk of unpleasant technical surprises after the start of production, and a need for consideration of production in the earliest stages of design,

h) more predictable product quality, as quality is increasingly ensured by process definition and control rather than by testing,

i) a substantial increase in the sophistication and cost of production process instrumentation,

j) a decrease in requirements for hourly factory workers, but an increase in the need for engineers and other professional workers,

k) an increase in the optimal size of enterprises engaged in the integration of all the elements of a product and the related services, with a concurrent increase in the number and the agility of small, technologically specialized firms providing industrial products and services in support of the "systems" firm.

This "new style" of engineering is, in fact, quite familiar in chemical engineering, where production processes are continuous and depend directly on chemical phenomena. The latest news is that the same advantages are increasingly available to the mechanical manufacturing of discreet physical parts. Its most conspicuous example is in microelectronics production.

2) Designer molecules and designer materials

Another consequence of the improved characterization of materials and the increased theoretical understanding of them is the ability to fabricate a manufactured material to meet specific functional requirements. It might, for example, be a composite material of fibers in a polymer matrix, or a laminate of different materials. By specializing the materials to specific applications, the latitude available for the product design is expanded, and superior solutions can be found.

Similar results are now being achieved by organic chemists and molecular biologists, whose knowledge of complex molecular structures has reached the point where new molecules can be deliberately constructed to meet specific chemical performance requirements.

These trends create markets for high value-added specialty materials, increase the product variety, and require specialized production facilities in which keeping costs low is a serious challenge. But they will also steadily impact the attractiveness of many natural materials, metals and metal alloys.

In so doing, they tend to uncouple the strategic industrial opportunities of nations from their natural resource base. It is therefore less certain from which quarter competition may be expected in world trade, and what the effect may be on commodity prices, already depressed by the substitution of manufactured materials.

3) Low technology and "bubble-up" engineering

The trends I have been describing are most commonly found in technologically intensive ("high-tech") industries. But they are also applicable in less sophisticated production where costs must be kept very low. Success in the manufacture of low-cost products comes from the appropriate use of advanced science applied in untraditional ways.

This may be illustrated by the example of digital electronics in consumer products. Many inexpensive products in consumer electronics (for example, the compact audio disc, or automatic cameras) use quite sophisticated components and mechanisms derived from recent, leading-edge science. However, the risk and cost are kept low by initially designing the use of the new, less well understood components with extreme conservatism. Once in production, the design parameters are progressively tightened, resulting in growing product function and market attractiveness with minimal increases in costs.

This mode of development, pioneered by Japanese engineers, differs from the traditional "trickle-down" approach, in which the first introduction of a new technology is in an application calling for a very sophisticated function, whose value supports the high costs of the early use of a new technology at its full potential. Then in the course of time, process yields are improved in production, and with

economies of scale from larger volumes of production, costs are reduced. In this "trickle-down" model, the low-cost consumer product
would be the last to use the new technology.

This was, for example, the history of the video tape, which was first
introduced by the American Amplex Corporation in a 3/4 inch industrial
format for use in television stations and recording studios. On the
other hand, thin-film transistor liquid-crystal display screens -
which have great potential in portable computers - were first introduced in Japanese portable TV sets.

"Trickle-down" diffusion of new technology is the common form of exploitation for civilian purposes of advanced military technology. To
the extent that "bubble-up" offers a superior alternative, the stimulation of civilian sector innovation must find its roots outside the
traditional defense-driven technical development.

This capability is also a by-product of the merging of engineering
with science, as mentioned above. It also illustrates the breakdown
of the "linear" innovation model, since the "trickle-up" approach to
engineering requires that new technology be introduced into production
during the early phases of product design.

4) Engineering for human use

As the technology content of human artifacts increases, for example
with the addition of microprocessor control of function, the improvement of function and efficiency may be achieved at the expense of ease
of use and ease of learning. Many people have strongly negative
feelings about the "dehumanization" of their lives when services become more impersonal and the tools they use every day become more
complex and mentally demanding.

Yet the increase in capability of human tools and the more economical
delivery of human services is a requirement for productivity growth,
without which standards of life cannot be improved.

The challenge is to make advances in human factors through engineering and bioengineering. Progress is being made, based on advances in the cognitive, computer and systems sciences.

This trend is most important in the computer industry, where the software carries the most important attributes of the way the system presents itself to the user (the "user interface"). Software is the most rapidly growing and most profitable sector of the information industry. A successful marriage of software engineering to human-factors psychology and human-factors engineering can open up a new era of growth in the information industry.

5) Engineering standards

Engineering standards will assume increasing importance in the next decade for the following reasons:

a) The science of measurement makes possible quantitative tests for performance-based specifications, so that high-quality standards become valuable assets in any system design.

b) Standardization that brings economies of scale from an aggregation of markets may accelerate a healthy growth of new technologies.

c) However, standardization can also be used as a non-tariff barrier to trade and may, if introduced prematurely or incorrectly, prevent the discovery of optimum technical choices.

d) Governments are increasingly involving themselves in the standardization of industrial products and services, and need to appreciate that the most successful standardization process is one that is driven by technically competent representatives of users of the technology, working with the manufacturers.

e) Standards development is a technically sophisticated task, for it requires an accurate, meaningful test development - a particularly challenging task when performance standards are sought. The use of performance standards instead of design standards increases the incentive for innovation while still protecting quality and safety. This activity will in future require a larger share of highly skilled technical people.

6) Engineering Education

Engineering education, and the relation of academic engineering to industry, will be the focus of change in a number of countries, but in different ways. Japanese university education is relatively weak in comparison to a rigorous pre-university education and a very effective system of training within industrial enterprises. The prime thrust will be a strong focus on basic science to balance the existing skill in applied research and manufacturing engineering.

American engineering education, which is superb in engineering science, must focus on training for "downstream" areas of engineering: design, process development, and production. Over 50 U.S. engineering schools have established new curricula in manufacturing systems engineering in the last seven years; this trend will accelerate and will have a positive effect on industry during the next decade.

German engineering is already strong, but primarily in the area of industrial capital goods. The need here is to improve the adaptation to end-user and consumer markets.

7) Intellectual Property

When product designs and production process controls are codified in computer programs, the information in these programs takes on exceptional value. The algorithms through which new molecules are designed-to-order and new materials are produced to meet specific application needs, reflect the costs of a great body of research and the value of a great variety of production units. If computer software can be devised that permits a naive user to master an information system and use it effectively without frustration, that software will be in exceptional demand. Its developer will wish to have that asset protected.

Finally, we now have the phenomenon not only of "designer molecules" and "designer materials", but even of "designer animals". Harvard University was issued a patent in 1988 for a new variety of mouse, created by genetic engineering for use in research on breast cancer. Under U.S. patent law, this permits the "inventor" of the mouse to charge a royalty in return for the privilege of breeding such mice, to be paid on each mouse replicated by the mice themselves.

All of these trends increase the cost of research in relation to production, and motivate their creators to seek assurance that their work is not taken by others without fair compensation. They reflect the shift of value added from raw materials and labor to facilities, instrumentation and knowledge.

Information (intellectual property assets) is becoming an indispensable element of competitive industrialization. As the factory was the key asset in the 19th century industrial revolution, adding value to raw materials, today intellectual assets provide the value added to production facilities.

Developing countries are quite conscious of this trend and are concerned about how they can assure access to this source of value added at a reasonable price, at a time when a worldwide collapse of commodity prices signals the end of an era when a healthy economy could be based on raw materials only. (Petroleum, the notable exception, has also lost much of its economic power with the world oil surplus depressing prices.) Refusal to participate in the international system for intellectual property rights protection may be frustrated, however, if the inventors shift their strategy for protection from patents and copyrights to trade secrets. World economic growth would be the loser in this event.

8) Interdependence from global environmental change

Environmental science now provides increasingly reliable conclusions about the effects of human activity on the biosphere. This progress results from the integrated study of the atmosphere, the oceans, the morphology of the planet, and its energy source, the sun. The average temperature of the earth's surface and atmosphere, the density of the ozone, which regulates the amount of ultraviolet light reaching the earth's surface, the effect of deforestation on the conversion of carbon dioxide into oxygen, and the production of acid rain from the burning of sulfur-bearing coal, can all be evaluated with increasing reliability on the basis of average global effects.

What science has much greater difficulty in doing, and may not be able to do until it is too late to reverse the effects of human activity, is to determine what regions of the planet these phenomena will affect most, or even the nature of the effect. Taking the greenhouse effect on atmospheric temperature, for example, some regions may experience an improved climate for agriculture; others may suffer desertification, with very serious consequences.

How will nation states find the motivation to cooperate to reduce these hazards, when only the collective effect on the planet can be predicted? The "tragedy of the commons" is serious enough without gross uncertainty about how small or how rich the commons may be.

9) Requirements for rapid global scientific progress

Finally, it should be observed that all these trends have directly or indirectly come from the progress of science. Scientific progress has come from four sources: quality education, a free and well-motivated spirit of inquiry, increasingly complex and expensive research equipment, and an open system of world communication in non-propriety, non-military research.

Education is a domestic matter for all countries. However, students in higher education may benefit from exposure to different schools of thought and styles of research by studying abroad. There are about one million students studying outside their home countries today. Approximately one third are in the United States. This sharing of knowledge and stimulation of talent provides the basis for the rapid diffusion of technology in the increasingly interdependent world economy.

It should be noted that many countries, including China, have adopted policies of encouraging study abroad as a means of investing in strategic capabilities in science and engineering. The United States has also used this strategy to advantage. The U.S. built its very successful scientific performance on sending its students to Germany in the 1920s, and on receiving scientists fleeing Hitler in the 1930s.

While such a strategy has its financial and potential political costs, it can be a powerful force for stimulation and renewal. The requirement for its success is, evidently, the availability of suitable jobs back home where the newly acquired skills can be fully utilized. Matching people to those jobs on the scale of China is a difficult task, especially in the absence of a truly free market for scientific talent and of the unrestricted mobility of scientists.

High motivation and high creativity and productivity for scientific progress are common to all societies that respect the autonomy of the scientist and encourage the twin motivations of intellectual curiosity and a contribution to the well-being of society. The pressures for accelerating the contribution to economic growth, felt keenly by scientists in the U.K., Japan, the U.S. and China, are a proper reflection of the increased value of science to engineering and to technology, but they also threaten the power of science to find the unexpected if the pressures are too strong.

All nations are also experiencing difficulty in adequately financing the facilities and equipment needs of modern science. Instrumentation development plays a particularly critical role in linking the discovery of new phenomena (analysis) to the control of new industrial processes (synthesis). As costs rise, scientists in all countries are learning to share expensive instrumentation.

Nations must learn to do the same thing, to share the development and use, for pure science, of the most powerful new instruments for discovery: optical telescopes, atomic particle accelerators, oceanographic ships, space science facilities, etc. A start has been made, but in the next ten years the cost of the greatest of such facilities will have risen to the point that even the richest nations will not be able to proceed alone. How the sharing will come about, I do not know. But I am confident that it will, for it must.

Finally, the encouragement of international discourse on fundamental science is necessary both for the healthy progress of world science and for each nation's scientists to stay abreast of their colleagues elsewhere. The recent history of discoveries in high temperature superconductivity should teach us that lesson. Scientists in the U.S., China and Japan almost simultaneously made the big leap to higher temperatures when they learned about the original Swiss discovery.

Nations must not let their legitimate concern about industrial or defense secrets, or the protaction of intellectual property, lead to a disruption of the open paths of communication in science and in basic engineering. The Japanese are worried that the U.S. reaction to Japanese competition in technology will result in U.S. "intellectual protectionism" to reduce intellectual "exports", even if America avoids more conventional protectionist barriers to imports. This danger will greatly lessened by the tenacity with which American universities defend their academic freedom.

Some specific areas of probable rapid progress in the next decade

1) Microelectronics

The aproximately 25%/year rate of growth of productivity will continue
to 1995, slowing down afterward. Current sets of industry standard
chips will give way to libraries of standard design modules, to be
combined with denser, larger chips of the future. Packaging, power
distribution and inter-processor high-speed communication will become
the challenging areas of computer hardware development. Consumer and
low-requirement industrial applications of microelectronics will domi-
nate the volumes of integrated circuit chips and will drive low cost
packaging and cabling development.

The supporting technologies will continue to rise in complexity, and
hence in capital cost. To compensate for these costs, larger rates of
production are required, implying both increased price competition and
increased industrial concentration.

2) Computers

Workstations, which are evolving to have a much higher computing power
than personal computers, will erode the demand for the present mid-
range products. By the mid-90s, workstations of a computational power
equal to today's supercomputers will be common. Parallel processing
will be well established in computers for science and engineering applications, permitting great extensions in processing power from today's supercomputer. These trends will support the extended use of
simulation and modelling discussed above.

3) Artificial intelligence

AI will be established as an important new application development
tool, but will not create separate markets or demand special hardware.
AI applications will generally be integrated into other application

systems. They will, however, bring the providers of information (today's publishers) together with the information processing and distribution industry to create a new range of information service industries.

4) Software engineering

There will be few technological surprises in software engineering; several versions of the UNIX operating system will gain acceptance as a common environment for application interchange.

5) Telecommunications

National telecommunications services are all going digital. Optical communications will displace satellites in all civilian applications except (a) intercontinental, (b) broadcasting, and (c) mobile and remote locations. Switched broadband networks are advancing very rapidly.

6) Materials

Engineered materials, especially composites, continue to displace natural materials and metals. Such materials are necessary for advanced electronic applications. Ceramic engines will be introduced into trucks using adiabatic diesel engines.

High temperature superconductivity (HTSC) will make a steady scientific advance, but major applications will be evolutionary and slow in coming. Chinese scientists were among the first discoverers of HTSC, and should continue to stay in the forefront of basic and applied research. Significant industrial and military applications are still some years away.

However, the cryogenic cooling of computers will be introduced in order to handle the heat dissipation from the very dense microelectric devices expected in the mid-1990s. Once computers are cooled in liquid nitrogen (to a temperature of $77°K$), there will be an opportunity for HTSC applications. Major engineering uses in transportation will be delayed until materials engineering brings competitive costs.

7) Measurement and characterization

Precision of measurement has improved by 10,000 times in the last 50 years. New tools, especially the laser, permit high speed, accurate, real-time control. The characterization of materials and processes makes possible modelling and simulation (for design) and production automation and control (for manufacturing). Materials science and engineering will continue to be the most important technical base for industrial innovation and productivity growth.

8) Genetic engineering

In agriculture, animal husbandry and improved natural materials, genetic engineering will have a broader economic impact than in pharmaceuticals because of delays for testing before human use. However, breakthroughs in immunology could enable pharmaceuticals to deal with many varieties of cancer. If this happens, there will be an extraordinary growth in the industry.

9) Unconventional energy sources

After a number of years of disappointment over the economics of alternative energy sources such as solar and wind energy, there may be some substantial changes "in the wind". New solid state electronics capable of a very efficient conversion of direct current into variable frequency alternating current with very carefully shaped wave forms, may

make windmill farms very much cheaper and competitive with other sources of energy. The same technology may also permit new forms of ultra
high speed motors.

Strategic choices

1) Multinational enterprises

European countries generally are responding to the expected EC market
integration in 1992 by looking beyond the support of national "cham-
pion" companies in protected domestic markets, to the formation of
joint ventures with international scope. Many of Japan's great industrial companies have not been true multinationals, in the sense that
they have concentrated on local production, exporting to world markets
out of Japan. The strong yen will accelerate overseas investment by
Japan in foreign industrial assets, especially in the U.S., thus broa-
dening the multinational character of Japanese firms. China may find
it necessary to explore more global ventures in the future in order to
take advantage of the "geographically distributed, vertical integration" that the new industrial technologies facilitate.

2) A global technological infrastructure

There is a clear requirement for a robust telecommunications infrastructure to take advantage of the science- and information-intensive
forms of engineering now emerging. This requires more than communications transmission facilities; technical professions and laboratories
need access not only to domestic science and engineering information
resources but also to the emerging international networks for sharing
scientific information and for scientific collaboration-at-a-distance
through computer networks. The most important of these networks,
called by some the World Univesity Network (WUNET), comprises the
European Academic Network (EARN) in Europe, the BITNET in the U.S.,
Northnet in Canada, and participating networks emerging in Japan.

The connection of such computer networks to industrializing and under-developed countries is a matter of great importance to their ability to participate in the global technical revolution. Understanding how this has developed and how participation can be facilitated, is an important issue.

3) A dynamic industrial strategy

As the high-technology segment of manufacturing begins to take advantage of the technological capabilities described above, a great flexibility is introduced into industrial strategy - indeed compelled by the dynamic optimization of business opportunities in other countries. This dynamism requires not only responsiveness to fast-changing markets, but an industrial structure capable of a great deal of agility. The generation of small- to medium-sized enterprises specializing in technology for industrial use is a key requirement for such an economy. Worldwide, such firms occupy technology "niches", and even when quite small, ofen have half their business from foreign buyers.

4) The globalization of service markets

Technology has also accelerated the globalization of service markets. In the Caribbean and in Asia (Taiwan, for example), there are rapidly growing enterprises using telecommunications, computers and well trained, inexpensive labor to provide "off-shore" services to foreign customers. This trend is also a part of the geographic dispersion of vertical integration, and may offer opportunities for China. Technology transfer itself will become an increasingly important part of the trade in services.

CHINA AND THE WORLD IN THE NINETIES

Trends in new technologies and their implications for China toward the 1990s

Keichi Oshima
Taiso Yakusiji
Technova Inc., Tokyo

I. Global trends in new technologies

Over the past fifteen years, the world economy has been subjected to a series of crises which have led to numerous domestic and international disturbances. Several major trends are at work behind this evolution of the world economy: the increased interdependence of national economies, the emergence of a multi-polar world, an increase in the welfare gap between the different regions and, most notably, the advent of a technological revolution.

This revolution is not just another of the periodic waves of technical change that have marked the progress of industrial society since its origins about two centuries ago. Today's emerging technologies, which are leading the world toward a services society dominated by information and communications, have within them the capacity to help solve many of the problems besetting society; but in order to fulfil their potential, they must be oriented toward the achievement of this goal by a global strategic vision and concerted efforts. This vision can be succinctly described as follows.

First, a closer look at current trends in technological development shows that what is happening today is a fundamental <u>paradigm shift</u>, not a mere incremental change in new technology, and that it will eventually lead to worldwide irreversible structural changes in social, economic and even political institutions. Examples are numerous. In particular, the paradigm shift can be observed in such technological areas as microelectronics, biotechnology and, most important, in information technology.

Europe-Asia-Pacific Studies in Economy and Technology
Leuenberger (Ed.) From Technology Transfer
to Technology Management in China
© Springer-Verlag Berlin Heidelberg 1990

Second, in the realm of industrial technology, the new technological paradigm realizes entirely new international industrial structures through what is called simultaneous off-shore production and procurement. Until recently, industrial production and procurement were restricted to local sourcing: production is performed at a particular industrial site where all productive resources are locally accessible. However, a new quantum leap in telecommunication technology has lifted such geographical constraints.

For example, new technologies need only a limited amount of qualified manpower locally, while quality control techniques, such as the just-in-time system, is widely diffused so that the rest of productive manpower can be obtained worldwide wherever manpower is cheap.

The implication of such global production and procurement is striking. First, national boundaries are becoming more and more obscure. Second, today's borderless economy is not the result of global investments only, but equally accelerated by the globalization effect of technology. Since the globalization of technology is moving ahead with a law of movement different from that for investment and monetary affairs, technological globalization sometimes moves faster than financial globalization, so that adjustments through exchange rates cannot always guarantee a stabilizing effect on world economic irregularities.

Third, the rapid progress of technology will change our notion of nation states. One reason is that, since the private sector is becoming a conspicuous active force in the field of technological transactions, there will be an increasing dissociation between the interests of transnational industrial groups and national governments; international technological negotiation will therefore take place between governments and companies. This trend is on the increase, which is why we are looking to LDSs and NICs for technological transfer.

Fourth, and in relation to the above third point, today's science and technology are becoming dual in nature. In other words, the international importance of science and technology is shifting from an economic to a political dimension. The recent deep anxiety about the impact of high technology on defense matters has been accelerating this development. In fact it is undeniable that some civilian technologies have an immediate impact on the advancement of military technology, software technology being particularly crucial.

Next, let me touch on some salient characteristics of today's four high-technology areas, namely information technology, bio-technology, the aerospace industry, and the civilian nuclear industry.

Information technology, which encompasses the full range from semiconductors to telecommunications, from computers and software to plant and office automation, and from professional to consumer electronics, are without doubt generating a change in the socio-technosphere of present-day society.

As far as bio-technology is concerned, it will have to be seen, at least for the next two decades, as an imminent technological breakthrough which will mainly affect the health sector, agriculture and related food industries, chemicals, energy, and the environment. Finally, while it can be seen that progress in the aerospace sector combines both radical and marginal innovations, the nuclear adventure we are witnessing at present is an obviously radical innovation with respect to electrical power and to energy generation.

II. Implications for China in the 1990s

Given the above discussion of the global trends in new technologies, I would like to move to some of the issues directly related to the China of the 1990s. It is quite possible that in the next few years, the international politico-economic scene will be subject to much more drastic changes than science and technology. By the end of the 1990s, however, new global trends affecting such new technologies as information technology and bio-technology will have made the conventional heavy manufacturing industries quite obsolete.

In this respect, China is faced with two choices. One is technology transfer, the other a kind of state-backed catch-up policy. The two may well complement each other, so that both choices could be taken simultaneously. Decisions in this field must be taken with caution in any case, since an overhasty approach might result in a vast shortage of foreign currency reserves.

In the 1990s, a global transfer of manpower and goods may be expected and, as pointed out above, production will be taken up on a global scale after a search for the best manufacturing sites available world-wide. This implies that developing countries, including China, could take advantage of exploiting technical know-how through middle techno-logies established on their territories, always provided that they can dispose of adequate infrastructures (including manpower training pro-grams) to integrate such technologies.

A crucial constraint in this respect will be the relationships between the state and the companies. In my view, the productive activities of companies would best be promoted through the establishment of state-owned holding companies that would supply investment capital for the formation of joint ventures. In our study of the Euro-Japanese High Technology Cooperation, one notable conclusion is that the best scheme of technological cooperation in the field of high technology is a joint venture. The Chinese government would therefore be well advised to provide such joint ventures with financial aid.

One last point which would have implications for China is the rela-tionship between high technology and employment. It seems to me that the reason why no serious labor problems have so far occurred in Japan may well be that there are no class distinctions in Japan. It may be noted, for instance, that most Japanese companies, both at home and overseas, have adopted the common lunch room system where executives and blue-collar workers share the same tables. In this way, new manu-facturing technology does not meet with strong labor resistance.

It is therefore advisable for China to embark on a social study focus-sing on the man/machine interface from a social and structural per-spective. Such a study should pursue new dynamisms of a social change that is different from classic cases. In any case, civilian technologies such as labor-saving technologies would not be implemented unless incentives can be clearly identified.

III. Some strategies open to China

In general, China and other developing nations are in a favorable position to welcome in the new era of high technology, in that such technology does not result in huge replacement costs. In fact, such costs are a serious burden for the industrialized countries, which have to face the renewal of existing industrial structures before the equipment is completely depreciated.

This advantage, however, can only be put to practical use if China itself makes the following efforts:

· To establish modern R&D institutes and invite a large number of first-rate research staff from abroad. Laboratories of this kind would have to include applied production research where non-production fields such as software and quality control technologies would receive a great deal of attention. The hardware-oriented labs should be located close to production facilities. It might be difficult for each company to finance such labs, so that some kind of state support would have to be envisaged.

· To implement an intersectoral fusion of defence and civilian industries. Naturally, some problems relating to official secrecy would remain in this respect. If the government recognizes, however, that economic gains exceed military losses, efforts should be made to accelerate such a fusion.

· To draw up a clear picture of the economic marketplace. To put it bluntly, I would think that priority should be placed on domestic markets first and on international ones later. In other words, it would be necessary to close domestic markets until China is capable of competing internationally. The drawbacks of this policy could be compensated for by various industrial sectors exporting a variety of goods. To think that China will be able to acquire enough strength to compete in international high technology markets would be to err on the side of optimism.

To found prestigious high-technology universities in China. History clearly shows that other advanced nations' industrial and technological competitiveness was based on the establishment of the world's best high-technology universities, such as Napoleon's Ecole Politechnique, Germany's Technische Hochschulen, America's MIT and Caltech, and many others.

CHINA'S STRATEGY FOR AGRICULTURAL DEVELOPMENT
IN THE 1990s

He Kang
Ministry of Agriculture
The People's Republic of China

In accordance with the decision of the Central Committee of the Communist Party and the State Council, our economic development has been carried out in three main steps. The first step was to double the gross national product of 1980 and to solve the problem of food and clothing for our people. This task has been largely achieved. The second step is to double the GNP again by the end of the century, thus enabling our people to lead fairly comfortable lives. The third step is to reach the per capita GNP level of moderately developed countries by the middle of next century. This means that modernization will basically have been accomplished and that our people will have begun to enjoy a relatively affluent way of life. The most important task at present is to make a success of the second step, in which a correct strategy for agricultural development has a direct bearing on the overall economic structure. To this end, we must pay attention to the following aspects.

1. With a view to working out a proper strategy for agricultural development, we must first of all analyse the supply and demand relationship of the main farm products according to national conditions. The growth of our demand for and supply of farm products by the end of the century will largely depend on three factors:

i. The increasing population will be a great pressure on the demand for food grain. By the end of the century, as the country's total population grows to 1.25 billion, the total demand for grain will reach 500 million tons, based on a per capita consumption of 400 kilograms. Should this level rise to 425 kilograms, the total grain needed will be 530 million tons. It will be necessary to increase the

Europe-Asia-Pacific Studies in Economy and Technology
Leuenberger (Ed.) From Technology Transfer
to Technology Management in China
© Springer-Verlag Berlin Heidelberg 1990

basic total grain production of 400 million tons by an annual average of 7.7 to 10 million tons, corresponding to average growth rates of between 1.7 and 2.2%.

ii. The second factor is a continuous increase in people's purchasing power, which will result in a great demand for meat, milk and eggs. It is estimated that the per capita purchasing power will increase from 345 yuan in 1986 to 731 yuan in 2000, and that the amount of money that can be used to buy food will rise from 190 yuan to 358 yuan during the same period. This amount is solely based on the consumer's quota of 400 kilograms of grain plus supplementary food. Given a higher purchasing power, people will naturally want to eat better and will be the first to consume more animal foodstuffs. Again, the growth rate of animal foods depends on the production capacity of grain and feed, so that a consumer's per capita quota of grain plus supplementary food is at least 425 kilograms.

iii. The third factor is a big increase in the demand for farm products due to the development of industry and foreign trade. 70% of China's light and textile industries depend on agriculture, which is why agricultural products and the processing of agricultural products occupy a decisive position in exports and in attracting foreign exchange: agricultural products account for 44.4% of the total exchange proceeds. There will be a greater expansion of foreign trade in the years to come. Besides, many of the exported agricultural products are either required in large quantities on the domestic market, or they fall short of supply.

Moreover, we have to take into account our natural resources, especially the low ratio of per capita arable land, which restricts agricultural development. The ratio of per capita farmland has dropped to 1.36 mu and is likely to drop further since more land is used for industrial buildings, roads and houses as the population increases. Other material resources, such as grassland and water surfaces suited for cultivation, are also limited. In the face of these constrained conditions, our objectives for agricultural development must be set to meet increasing public demand as well as the actual possibility of an increase in agricultural production.

2. In consideration of China's present-day conditions, our goals for agricultural development by the year 2000 must be these:

i. A sustained growth of the rural economy and of the farmer's income must be based on constantly improving economic results. The total social output value in the rural areas will reach about 1,900bn yuan, of which agriculture will account for more than 540bn yuan. A person living in the countryside will have an average annual income of 800 yuan; the majority of the rural population will achieve a relatively comfortable standard of living.

ii. The major agricultural products will basically provide the population as a whole with a good livelihood, as well as satisfying the needs of a national economic development. By the end of the century, per capita consumption of agricultural products will reach 400-425kg grain, about 25kg meat and 14kg fish and seafood. It is necessary to make some adjustments to improve the diet, with a proper increase in the proportion of animal foodstuffs. A persons's average daily food intake will supply 2,780 calories, 77g protein and 62g fat. While public demand for cotton fabrics is fundamentally met, the consumption of chemical fiber, cotton fabrics mixed with silk, ramie, linen, etc., as well as woolen goods, silk and satin, should be increased accordingly. The export of agricultural products is to be expanded properly, with an emphasis on increasing the share of processed products in overall agricultural exports; processed products should then make a considerable contribution to China's foreign exchange earnings.

iii. Making great strides in promoting the level of agricultural modernization: the aggregate power used by the country's farming machinery will reach 500 million horse power. The key operations in agriculture, forestry, animal husbandry and fishery should mainly be done by machinery. The total farming area under tractor ploughing and sowing will reach 70% and 60% respectively, and 80% of farmland irrigation will be done mechanically or electrically. The quantity of chemical fertilizers applied will reach about 150 million tons, up 80% from the present level. The overall consumption of electricity, gasoline, diesel oil, insecticides and plastic sheets will increase faster in rural areas. Agricultural modernization will initially be achieved in economically developed areas.

iv. Rural labor forces will be transferred to non-agricultural sectors according to definite plans. By the end of the century, a 170-200 million strong workforce employed in village and township enterprises and other non-agricultural jobs, i.e. secondary and tertiary sectors, will account for over 50% of overall rural manpower. Once rural markets are built up, towns and small towns will become links connecting cities with the countryside.

3. The strategic alternative of our agricultural development

Agriculture is a basic industry that affects the overall situation of the national economy. The development of industry as a whole and of agriculture-related industries in particular depends to a great extent on how much grain and raw materials farmers are able to supply; it also depends on the handling of the commodity market. China is a developing country with a low level of public production forces; its socialist society is still in its primary stage. For such a country, the steady growth of agriculture is not only an important condition for a sustained and coordinated development of the national economy, but also an important guarantee for social stability and unity. On the other hand, China has a huge population and scarce material resources, which is apt to make an adequate supply of agricultural products a long-term problem. In order to solve this problem, however, we cannot place our hopes in a constant import of grain; rather, we shall have to keep on improving our agricultural productive resources.

Under the dual pressure of an expanded public demand for agricultural products and the relative scarcity of our natural resources being increasingly noticeable, the only choice of our strategy for agricultural development is to implement intensive farming so as to step up productivity per resource unit and thus maximize our total agricultural production. Before the growth of our population reaches its climax, either this century or even in the middle of the next, it must be our fundamental policy to take every opportunity to exploit, utilize, and economize on arable land, grassland and cultivable water surfaces. Moreover, we must institute and develop land-saving crop farming as well as grain-saving animal and aquatic cultivation. Agriculture must be in accordance with our natural situation, so that it will not only be instrumental in improving the composition of a diet with Chinese characteristics, but will also help to form and gradually perfect a system of production and technology capable of producing plants and animals with a high and stable yield, high quality and productivity, but low costs. Our policies on agriculture, industry and technology must be in favor of a successful implementation of this strategy for agricultural development. It is to this end that reform efforts must be persistently intensified and the following six strategic measures firmly carried out.

i. Bringing about the integration of the urban and rural economies through an adjustment of the industrial structure

In order to press on with a constant development of agriculture, rural areas must be industrialized, their populations urbanized, and secondary and tertiary industries developed in line with local conditions, making full use of the abundant human resources available in rural areas; in this way, urban and rural economies are promoted simultaneously. We place our hopes of agricultural modernization in village and township enterprises comprising industry, transport, construction, trade, and services. The steady development of such enterprises and an accelerated transfer of rural labor forces to non-agricultural industries are the preconditions for an improvement in agricultural productivity. Village and township enterprises in coastal areas should gradually enter into the orbit of the export-oriented economy, in an attempt to become involved in international exchange and competition. The close economic relationship between such enterprises and agriculture should be maintained, since agriculture is a production basis supplying products required by these enterprises as well as by their personnel. It is therefore necessary to keep on perfecting the various ways of making industry support and build up agriculture, and to promote the material and technical foundation of agriculture as well as its ability to remain self-sustained. In this way, the secondary and tertiary sectors in rural areas can be developed in coordination.

ii. Creating the right economic relations to improve the operating mechanisms of the commodity economy.

Farm prices must be rationalized. According to the principle of exchange at equal value, the differentials now prevalent in the exchange of industrial products must gradually be narrowed so that the producer's comparative advantages improve and farmers are likely to grow more produce. We should set up rational parity prices for products as quickly as possible, and keep an approximate balance between the comparative advantages of the various products; this would promote their coordinate development. In the basic interest of consumers, we must plan to straighten out the buying and selling prices of the main agricultural products, to revitalize the managerial abilities in the chain of commodity circulation, and to aim at an increase in consumption appropriate to the development of production. Efforts should be made to develop new types of integration of production, supply and marketing; subsequently, domestic and international trade must be integrated. An organical association can first be tested in the area of non-staple foods, with the help of conditions created in the process of the reform of political and economic structures aimed at serving increasing public demand more effectively.

iii. Increasing investment in agriculture for its further development.

We must not only preserve present production capacities but create new ones through an increase in productivity. The state's investment in agriculture should grow in proportion to national strength, but this is not enough: local authorities, too, should allocate more financial resources to agriculture. Moreover, we should adopt preferential policies of investment in order to encourage the farmers themselves, village and township enterprises, and large and medium industrial and mining companies to invest more capital in agriculture. At the same time, we must actively, if gradually, press on with farm mechanization in accordance with local conditions. We should secure and increase the production and supply of such materials as chemical fertilizers, pesticides, diesel oil, plastic sheets and farm machinery compatible with the appropriate agricultural development.

iv. Improving specialization in farming, creating a large-scale management structure, and strengthening internal forces in agricultural development.

No time should be lost in the development of farming households specialized in crop farming, animal breeding and fish raising in areas where village and township enterprises are relatively well-developed and where the land/laborer ratio is higher. In those areas, land productivity and the return rate of feed utilization, as well as labor productivity, must be improved. This should be done in a manner which ensures that economic returns will be of such a scale that those engaged in agriculture will have a slight comparative advantage over those working in the non-agricultural sectors; in this way, agricultural production can be continuously intensified and extended. The emphasis on the development of a fairly large-scale management structure is to create a requisite condition. We have to solve various problems that have emerged since the institution of contracted household responsibility, a system where remuneration is related to output, and which has been set up in order to establish and gradually perfect a public service. As long as the two-tier system of collective and farmer-household management is constantly improved, it will be possible to manage agriculture on a fairly large scale without any difficulties once the right conditions prevail. By the end of the century, large-scale management structures will basically have been implemented in areas where such conditions apply.

v. Improving the standards of agricultural science and technology
by promoting the quality of the workforce.

Regardless of other preconditions, the productivity of every resource
unit ultimately hinges on the standards of agricultural science and
technology. These, however, cannot be upgraded unless the general
quality of the agricultural workforce is improved. It is self-evident
that the development of intensive farming requires great efforts in
training better-qualified farmers. The reforms of higher and secondary education must include a forceful drive to develop adult education; also, education must gradually be made compulsory, and a multi-
level educational system must be created. At the same time, top priority must be given to making progress in science and technology with
regard to the development of agricultural production, to accelerating
the reform of the country's science and technology structure and of
its service system, and to delegating more power to institutes and
their faculties so that research results can be translated into productive resources as fast as possible. The proportionate effect of
science and technology on the growth rate of agricultural production
is expected to rise from the present 30% to 50-60% by the year 2000.

vi. Controlling the population growth and the use of farmland for
construction, accelerating the exploitation and utilization of
agricultural resources to protect per capita material
resources.

China's per capita agricultural resources are relatively scarce on
average; this constitutes the main limiting factor in our agricultural
production. In addition to improving productivity per resource unit
and exploiting the reserve material resources by every possible means,
we must regard the control of the population growth and the use of
farmland for construction as the most important long-term strategy in
the years to come. More effective supporting measures must be adopted
economically, administratively and legally in order to keep our total
population within 1.25 billion by the end of the century, with the emphasis on checking the birth rate of the rural population, which has
been in excess of the planned target. Meanwhile, the amount of land

used for the construction of village and township industries, roads and houses must remain lower than the amount of newly reclaimed farmland. The principle of no, or at least less, appropriation of good farmland for non-farming uses should be observed resolutely. The establishment of a multiple crop index and the speeding-up of the development of agricultural resources should ensure that our total amount of farmland will not be significantly reduced. Also, the decrease in per capita farmland should be kept within a range of 1.8%, the range set for the planned population growth rate. In this way, we can make sure that our farmland will basically meet the needs of our agricultural development when China's population reaches its peak in the middle of the next century.

SCIENTIFIC AND TECHNOLOGICAL PROGRESS
AND THE REVITALIZATION OF CHINA'S ECONOMY

Hu Ping
President
National Research Center
for Science and Technology Development

I

Human civilizations in both their material and intellectual spheres
have been shaped by advances in science and technology. The current
technological revolution is promoting social development so greatly
that it is leading to a new international dynamism and framework in
the world economy. Countries which are relatively highly developed
are likely to fall behind if they attach less importance to science
and technology (S&T) progress, while less developed economies may be
in a better position to seize the opportunities offered by a new and
wider range of technologies. Thus experiences and lessons drawn from
what determines the efforts of late-comers deserve our attention.

Having failed to seize the rare chances offered by technological revo-
lutions several times due to historical and social reasons, China
still remains relatively backward in its economic and social development. Despite several ambitious initiatives by the government to
raise productivity, such as the Big Leap Forward - a nationwide cam-
paign of the late 1950s - China attempted in vain to make a big thrust
in its economic development, for it failed to harness technology to
its economic development while advocating "hard work" and the costly
use of raw materials, equipment and human resources. For decades
China has followed an "extensive" development route, which has led to
little development of its technological capabilities. The technological level of manufacturing remains at a stage reached in the 1950s or
earlier by developed countries. The consumption of raw materials and
energy, however, is several times higher in Chinese manufacturing than
it is in developed economies, while the rate of value added by fixed

Europe-Asia-Pacific Studies in Economy and Technology
Leuenberger (Ed.) From Technology Transfer
to Technology Management in China
© Springer-Verlag Berlin Heidelberg 1990

assets and labor productivity is so low that it accounts for only a fraction of that of developed economies. China has one of the lowest rates of value added and labor productivity in the world economy. At the same time, the population growth is putting pressure on the availability of arable land and natural resources - which are still relatively scarce. Therefore if we remain with the current backward technological basis, it will become impossible to survive.

Since the Third Plenary Session of the 11th Party Central Committee, the Government has shifted its strategic development efforts toward increasing productivity in order to promote economic development through scientific and technological progress. This has been made a top priority in the S&T community. Some initial results have been achieved in this respect. However, for a country with a population of one billion, a persistent effort lasting several decades will be needed in order to realize a fundamental change. Only when industry and agriculture are developed on the basis of modern science and technology, can China develop a material base which is adapted to determine survival in the future. China is quite capable of making efforts in this area, and although its S&T capability is not as strong as that of most developed economies, it is strong in comparison to the developing countries and even some of the Asian NICs (Newly Industrializing Countries).

An integrated S&T system has already been established in China. There are about nine million S&T personnel who have been trained in a variety of disciplines, and some of their accomplishments have reached advanced levels internationally. China has established technological capabilities in the frontier areas of nuclear power, space and microelectronics. Unfortunately, it has been constrained by an ossified social system. The technological capabilities have been lying idle for so long that they have not been used to achieve economic profits. The objective now is to turn our technological advantages into economic ones. A Medium and Long Term Science and Technology Program is currently being defined with this as its goal. In this Program, the objectives and major policy measures for harnessing S&T to future economic and social development are detailed in various dimensions, such as: upgrading technology in existing industries, fostering new high-tech industries, developing modern agriculture, guiding social development through S&T. All this is bound to have a far-reaching impact on China's future development.

II

The battlefield on which S&T will be developed will be the transformation of traditional industry and the application of modern science and technologies to effect a vast increase in labor productivity. A relatively complete traditional industrial basis has been established, consisting of assets worth more than 600bn yuan. By 1986 the total annual industrial output value covered 73.6% of the total agricultural and industrial output value. The foreign exchange earned by industrial exports amounts to 64% of total foreign exchange earnings. Traditional industry will remain the leading sector in China's future economic development; only 20% of current industrial technology has reached the international levels of the 1970s. In most traditional industrial sectors, obsolete equipment is still in operation. Processes are backward, and raw material consumption is high, which contributes to poor output quality and low profits. Faced with such big challenges, the general goal of China is to transform existing industries and, by the year 2000, to have a technological capability in the main industries of a 1970/80 world level. The rate of achievement of international standards in key industries is to be raised by at least 50%.

This economic transformation with the use of S&T will require the establishment of priorities. In the current context of China, these priorities will be:
1) energy industries based on electricity;
2) raw materials, iron and steel, and synthetic materials;
3) electronics in the mechanical engineering of large-scale turn-key plants;
4) transport and telecommunication networks.

Painstaking efforts should be made to develop a complete system of technology and equipment, electronic information technology, and integrated mechanical and electrical technologies. Priority in technological transformation is given to industrial processing technology, raw materials, equipment, and intermediary products. Great efforts must be made with regard to the development and diffusion of appropriate advanced technologies.

While prompting technological progress in the main industrial sectors, we will remain concerned with technological progress in the rural areas and in numerous township enterprises. We will continue to make great efforts to implement the Spark Plan, and to provide S&T to poverty areas. The output value created by China's rural and township enterprises is remarkably high. This will significantly stimulate the commodity economy in China. Technology employed by such enterprises is generally backward and must be transformed through upgrading and the introduction of advanced technologies. The comparative advantage of rural areas is their cheap labor cost. As long as technology in rural areas is upgraded in order to produce labor- and technnology-intensive products, a niche may be found in the international market. With better transport and communication systems and with higher educational levels, the coastal areas are provided with more opportunities to develop outward-looking economies. A drive will be launched to foster rural development and agricultural modernization.

In the process of technological transformation, strategic importance will be attached to advanced technology from abroad. This will require four steps away from past experience. The first will be to move away from the importation of turn-key plants and production lines toward "soft" technologies. The second will be to move away from research efforts with regard to technology import programs at enterprise level toward joint efforts by research institutes, university and industry. The third will be to move away from the simple "use" of technology toward its absorption and toward innovation. The fourth will be to move away from one-way importation toward joint research efforts by industry and research institutes with foreign enterprises, both in manufacturing and in research products. This will mean that technology transfer may play a more substantial role in China's economic growth.

III

The development of agriculture is vital to China's survival and prosperity. Although a great deal of progress has been achieved in the past decade, agriculture still faces the great challenge of its rapid population growth and of pressures from arable land. Vital measures are needed to change China's agriculture by means of modernization using S&T.

The total industrial and agricultural output value for the year 2000 will have to be doubled in comparison with present production levels, and at least 500 billion kilograms of grain will have to be provided to achieve the national economic goals. A major change will have to occur in China's diet, so the supply of animal protein will also have to be doubled. Meanwhile the provision of various economic crops will also have to be introduced on a large scale. How will we reach these targets without further environmental deterioration? Apart from policy measures based on scientists' suggestions, technological measures must be taken. These include:

1) The transformation of land output to higher levels through the application of S&T. We are planning to transform 3 million mu of land in the Huang huai hai plain, which means that an additional 150 billion kilograms of grain will be produced.

2) To develop plough substitution technology by promoting agricultural processing, vegetable gardening and irrigated cultivation, and to increase highly intensive multiple crop systems.

3) To improve hybrid strains, optimize fertilizer compositions, and diffuse polythene technology, as well as water conservation technologies for purposes of irrigation. Our successes in this area indicate that polythene coverings alone may increase yields by 22-30%.

4) The establishment of a far-reaching agricultural strategy, aimed at raising protein levels by utilizing China's full grassland potential as well as hillsides, coastal areas and marine resources. This will help change the conventional practice of living on grain alone and reduce the reliance on only this crop for raising pigs. An increase in aquatic cultivation and fisheries will mean that food sources will become more varied.

5) To restructure agriculture in order to achieve economies of scale by integrating separate pieces of land so as to create the necessary conditions for automation, thus speeding up the process of transforming the rural labor force into an industrial labor force. Hopefully, 200 million rural laborers will be transferred to the industrial and service sectors by the year 2000.

6) To strive to develop agriculture, especially in the coastal areas, which are capable of earning foreign exchange.

For Chinese agriculture to be competitive on an international level, all this will involve the development of labor- and technology-intensive products and the application of advanced technology. If these policies are implemented, however, China's agriculture may develop in a dramatic manner.

IV

At present, high technology has increasingly become the spearhead of international competition in scientific communities, in civilian and military technologies, and in economic affairs. A group of leading technologies emerging in the 1970s - microelectronics, biotechnology and space technologies among them - permeated, and were diffused in, the key sectors of the international economy. This has radically improved productivity, with concomitant changes in the world economic structure and in employment. This presents developing countries with a great challenge, but also with opportunities. If China does not consider the implication of the diffusion of these technologies, China will not be able to modernize and to reach its desired economic growth. It is not possible for China to develop capabilities in all these areas immediately, but it is both rewarding and possible to identify priorities and to concentrate efforts on key technologies and on specific technological breakthroughs.

The priorities selected by China include biotechnology, space technology, information technology, automation technologies, new energy technologies, and new materials.

The potential of high technology is based on its high value added, its pervasive efforts in the field of economical production in all sectors, and its promise to improve economic efficiency. The heavy investment effort for R&D activities in the field of high technology is meant to provide the foundation for an actual industry. Although the issue of high technology and its development has been studied extensively in China, fewer efforts have been made to develop and produce these technologies at enterprise level. We have not moved much further than research - at least not in the civilian sector. Applications of hi-tech have been confined to the military sector rather than spreading to other sectors of the economy. The result is that we have a relatively high level of research in the area of high technology compared with a low level of application in productive industry.

The development of high technology industries need not be dependent on the current level of technological capability in that field. In some applications, for example, China already has the necessary conditions for the development of high technology. Moreover, restructuring processes in the developed economies provide China with several opportunities to develop key applications. As long as we adopt a position of offering support and of encouraging the development of high technology in directions selected with a view to both the international markets and our own context of comparative advantages with regard to raw materials and low-cost skilled labor, we will be able to enter the international market and improve our role in the economic system. One important prerequisite for access to the international market is the selection of appropriate areas in which to attempt breakthroughs, the exercise of our comparative advantages, and the utilization of our existing capacities such as turn-key plants and intermediate products, in order to achieve economies of scale.

Based on previous experience, as well as on openings in the international market, China is in a good position to make breakthroughs and establish industries, in the relatively short term, in the field of information technologies based on microelectronics, in new materials, and in biotechnology. At present, the proportion of the value of high technology in the overall economy is very small - just under 1% at most. As long as we adopt appropriate policies, we can increase this proportion several times over by the year 2000. This would produce major changes in China's economic structure, and a new outlook would emerge in China's modernization process.

V

Science and technology are among the most powerful tools for promoting economic development, and the most powerful instruments for promoting the progress of civilization.

Modern science and technology have brought human society to a new stage of development and have created a new stage in civilization. Due to the influence of several thousands of years of feudalism and due to a history of self-reliance, many areas of Chinese society are still hampered by a lack of development and by benighted attitudes. This is not compatible with modern society. All these defects can be overcome through the development of science and technology. Thus one key principle is to transform backwardness in our society and to promote a health-improvment program through the application of theories, methodologies, and modern science and technology.

The most important task faced in the development of science and technology with a view to social betterment, is to control the population growth and to ensure good health for future generations. This will involve the use of science and technology in an attempt to change people's attitudes toward having children, including the popularization of eugenics and improved standards of child care. Scientific methodologies should be used to forecast and monitor population growth trends. The population growth must be compatible with such a sustained growth as is permitted by economic and resource constraints.

Another important task for S&T is to direct the developments in the field of the protection of the environment and the preservation of an ecological balance. With the development of the economy over the last 30 years, the quality of our environment has been rapidly deteriorating. The amount of atmospheric dust has increased many times over, water pollution has reached serious levels, there has been an increasing desertification of arable land, and we have a great deforestation problem. If we failed to change this trend resolutely, the future would be unthinkable. Therefore we must adhere to ecological regulations and formulate a national program for the environment in order to establish an environmental protection system and achieve a substantial improvement by the year 2000.

The third important task ahead of S&T is to improve the quality of life for people. We must use modern technology to reduce heavy and dangerous work and to improve working conditions. We must improve hygiene and health standards, control common diseases, and prolong life expectancy.

The fourth task is to change our stereotyped concept of living and to establish a new way of life. China will therefore try to introduce new knowledge into our ethical concepts and moral values, into our general and physical culture, in order to change the thoroughly backward feudalist concepts and conventional habits, advocate atheism and criticize theism, and establish new value concepts ways of life compatible with a socialist commodity economy, by criticizing idealism through materialism, and to promote the intellectual level of Chinese society to reach new heights.

The development of S&T, the promotion and revitalization of the economy, and the construction of modern socialist China, represent a great challenge to the Chinese people. In order to be equal to this challenge, China must work persistently hard through many generations, and the next few decades will be key ones for China's economic development.

The development of S&T is of decisive significance in determining whether China's economic development will be successful. The process of S&T development is not only an economic one. Rather, it will involve the vitality and creativity of our people. Science belongs to the future of society as a whole. Once science and technology are fully developed in China, the realization of modern socialism will have been achieved.

CHINA AND THE WORLD IN THE NINETIES

Deepening Reform for Technological Progress in China

Lin Zixin
Editor-in-Chief
Chairman
Science and Technology Daily
China

Since the founding of the new China, a relatively integrated industrial system has been established after nearly forty years' efforts. Industrial and agricultural production has developed rapidly, and scientific and technological standards have risen remarkably. By 1986, China's output of major industrial and agricultural products was the largest in the world: for example, cereal, meat, cotton, cloth and cement ranked first; coal, second; chemical fertilizer, third; steel, fourth; and crude oil, chemical fibers and electric energy, fifth. The number of natural scientists and technical personnel in China reached 8,253,100; the number of R&D institutes in natural science and technology affiliated to governmental organizations above county level was up to 5,271, employing 1.02 million staff, of whom 324,800 are scientists and engineers. The scale of these operations also put China among the first in the world.

Since China is a developing country, however, its present productive resources are still far behind those of developed capitalist countries.

We could probably draw a fairly practical conclusion by comparing China with the United States, since both China and the United States have vast territories and huge populations. The Chinese territory covers 9.6 million square kilometers as against the USA's 9.36 million square kilometers; the Chinese population by the end of 1986 was 1.5721 billion compared with 239.4 million in the USA at the end of 1985.

Europe-Asia-Pacific Studies in Economy and Technology
Leuenberger (Ed.) From Technology Transfer
to Technology Management in China
© Springer-Verlag Berlin Heidelberg 1990

Comparison of the Productive Resources of China and the USA

Items	China (1986)	The United States year	data
persons supplied per farmworker	3.4^1	1820	4.1
illiteracy rate	23.58% $(1982)^2$	1840	22%
per capita output of steel	49.38kg	1887	57kg
per capita electric energy	425kWh	1917	421kWh
railway operation mileage	52,500km	1863	53,400km
highway mileage	962,800km	1904	3,460,000km
telephones per 1,000 people	6.7	1897	7.1
proportion of labor 1st:2nd:3rd sectors	61:22:17	1870	52:24:24
population ration urban and rural areas	23.5:76.5 $(1983)^3$	1870	25.7:74.3
per capita GNP (at 1980 prices)	nearly $400	1869- 1878 ^{4}	$1,328

Notes:

1. The figure is based on 313,110,000 manual laborers in agriculture, forestry, husbandry, and water conservation.
2. If the number of illiterates and semi-illiterates is divided by the total number of people in the same age groups, the illiteracy rate will be 31.87%.
3. Considering that the term "township population" stands for the entire population of all areas under that township's jurisdiction, and considering that many new towns have been established since 1984, I have not quoted the 1986 statistics here, which would have been 41.4:58.6.
4. average value / year

The above comparison demonstrates that in terms of productive resources, China lags more than one century behind the United States. Obviously, this differs greatly from what we usually say, namely that
"the technology gap between China and the developed countries is about
ten to twenty or twenty to thirty years". The reason for this is that
with regard to the development of productive resources, a good grasp
of technology is certainly important, but the application and diffusion of technology are even more so. For example, today's technological level of the Chinese railways is very close to that of the developed countries, yet our railway mileage and the amount of rolling
stock does not match our land area; our present annual output of steel
is up to two thirds of the American output, yet our aggregate output
of steel from 1890 to 1985 is only 644 million tons, which is equiva-
lent to the American output of 604 million tons between 1966 and 1970.
The increase of social labor productivity is even more dependent on
the productive equipment used by each laborer. In 1986, the production equipment owned by each farming household in China was worth less
than 300 yuan (137 yuan for transport equipment, 17 yuan for agriculture, forestry, husbandry and fishery equipment, 21 yuan for industrial equipment, and 48 yuan for large and medium farming tools made
of iron and wood), while in 1984, each American farmworker owned mechanical equipment worth an average $34,000 - a difference of several
hundred times over.

All this shows that the development of productive resources does not
only need a good grasp of technology but also requires an input of capital and time. Moreover, the state of Chinese natural resources must
also be taken into account.

Considering its overall resources in cultivated land, forest areas,
grasslands, water and major mining reserves, China is certainly not
among those countries that lack resources. Its huge population, however, means that the per capita level of natural resources is usually
lower than the average world level.

Comparison of Natural Resources of China and the World (1985)

Items	China	World average
	(per capita level)	
territory (ha.)	0.92	2.80
cultivated land (ha.)	0.09	0.28
forest land (ha.)	0.11	0.85
grasslands (ha.)	0.30	0.65
annual river runoff (m^3)	2,473	10,050

Here, it is not only a matter of quantity. For example, the per capita annual river runoff in China is lower than the fresh water volume at present developed and utilized in the United States (2,600 cubic meters a head per year); indeed, the lack of water resources has become a factor restricting both rural and urban development. The per capita amount of cultivated and forest land is insufficient, and what is more, it can only be increased by a small margin because the remaining undeveloped land only makes up about 15% of the existing arable land and about 75% of the existing forest land. Besides, there is a quality problem. Of the Chinese territory, 33.3% is mountainous (compared with 10% in the USSR and 15% in the USA) and 26% high land, which adds up to nearly 60%. In existing cultivated land, mid- and low-yield fields amount to two thirds, high-yield fields merely to one third.

Thus productive resources are very backward, natural resources per capita rather low, the growth of the population fast, and capital accumulation slow: this is the severe reality we must face when we formulate a national science and technology development strategy for China. Even so, China is a country integrated into the modern world and must therefore take full consideration of the rapid development trends of the new worldwide technological revolution and adopt various appropriate new tech discoveries in good time. At once taking into account the national situation and facing the world, we are undertaking the dual task of concentrating on a traditional industrial revolution, while at the same time trying to catch up with the new worldwide technological revolution. How to deal with the two properly takes a great deal of learning. There are differing opinions, but it seems that the more practical strategies can be summed up as follows:

1. The most important task for science and technology (S&T) development: revitalizing the national economy. In order to make S&T the most important tool for national economic development, we must adjust the distribution of S&T resources, including military use / civilian use, research / development, scientific research and education / production, etc. Most S&T resources, and particularly the majority of the best qualified personnel, must be concentrated on serving economic construction.

2. A long-term strategic goal: closing the technology gap. Attention must be paid to closing the technology gap between China and the developed countries. Even more emphasis must be placed on closing the technology gap between individual regions, urban and rural areas, industry and agriculture, as well as large and medium or small enterprises in China. First of all, the focus should be on technology for mass production in the areas of agriculture, energy, raw materials, transport, communication and engineering. Technology development should be speeded up step by step and in a well-planned manner; in the meantime, we will have to put up with a certain amount of backwardness that is difficult to change for the time being.

3. The important principle in technology choice: to save capital and to conserve resources. This is quite different from western countries, which put stress on labor saving and which substitute a labor shortage with large amounts of capital and energy. The per capital level of natural resources in China makes it impossible to sustain a "throw-away economy"; neither can China provide every worker with any fairly large amount of capital within a short period of time. The selection and development of the kinds of technology appropriate to China's condition is an arduous task facing our scientific and technological experts.

4. The mass foundation of technological progress: a high number of skilled workers and technicians. Paying attention to training engineers is undoubtedly right. If, however, there is a shortage of skilled workers, any excellent design, state-of-the-art process or technical equipment cannot be realized as expected. In the long run, technological progress reflects the improvement of technical skill and the knowledge put to use by the workforce as a whole.

5. The key measures to close the technology gap: importing advanced technology from outside. While it is difficult to select optimal projects and raise the requisite amounts of foreign currency, it is even more difficult to digest and disseminate imported technology. There are several urgent problems which remain unsolved, among them the people's wish for technological progress, as well as the organization of the Chinese R&D resources to digest, disseminate and develop imported technology.

6. In-depth plans integrating the long and the short terms: actively to strengthen basic research. How long will such processes last, and how far will they go? These are controversial issues. Because of limitations on capital and human resources, decisions must be made as to whether certain programs are to be pursued or rejected. It appears that we should only focus on those projects which have the advantage of developing productive resources within ten to fifteen years; and even within those parameters, our capabilities and capacities will put certain restrictions on us with regard to "high-science" projects requiring large investments.

On the whole - choosing between science and technology, we should give the acceleration of technological progress priority, with special attention being paid to the dissemination and full utilization of technological achievements from all over the world, including high-tech discoveries geared to our own conditions. This would help us achieve a revolution in the traditional industries by the middle of next century, so that we would be able to catch up on our technological "shortfalls".

To envisage such a strategy may appear to be somewhat conservative and short-sighted, yet I regard it as quite a practical one which is even imperative if we want to accomplish the second step in China's economic development strategy. This second step is to double China's 1987 GNP by the end of this century, thus enabling our people to lead a fairly comfortable life. Also, this will still leave us sufficient leeway to increase our expectations once the time is ripe.

Let us look back on over forty years of Chinese effort. In the 1950s, the Chinese government called on the people to march toward science, and to catch up with and even surpass the world's advanced levels; China was the first country in the world to draw up a "Twelve-year Long-Term Plan for Science and Technology". In the 1960s, China defined the modernization of S&T as one of the four fundamental types of modernization. In the 1970s, China affirmed the status of S&T as productive resources, indeed as the keys to the fundamental types of modernization. Finally, in the 1980s, China advanced a policy maintaining that "economic construction must rely on science and technology", calling on people to lose no time in meeting the challenges of the new worldwide technological revolution and "to give first priority to the expansion of scientific, technological and educational undertakings so as to promote economic development through advances in science and technology as well as through an improvement in the quality of the workforce." This strong emphasis on S&T expressed by the Chinese government may not be quite as common in other countries, and it is due to this great support on the part of the government that the number of scientific and technological personnel has increased rapidly, and that great achievements have been reached, particularly noticeably in the high-technology sector of national defense. Owing to drawbacks in the S&T system, however, S&T has been divorced from production, R&D findings are difficult to use in production, and many as yet unsolved technical problems affecting production have been neglected for a long time. All this has had a strongly adverse effect on the improvement of productivity. In 1985, after the decision made by the Central Committee of the Communist Party and the State Council on the structural reform of S&T, the above-mentioned situation was changed in some of its aspects, but still not to our satisfaction.

How, then, can the reform be intensified in order to speed up technological progress? I think that a full use of the mechanisms of the
commodity economy is the key to it. The Chinese government has com-
missioned studies in this field during the last few years, and some
socialist and developing countries have also gained some promising
experience in this respect. From a preliminary point of view, attention should be paid to the following points:

1. To aim at markets both at home and abroad, to take economic results as the most important criteria, to measure the technological
level of production, and to give technology full play in commodity
production.

2. To use competitive mechanisms and the law of value in order to encourage engineers, technicians and shop-floor personnel to put their
efforts into improving product quality, into reducing the heavy consumption of materials, and into lowering the high production costs.

3. To form a tripartite cooperation system with a financial institution assuming the function of an intermediary which will not only be
able to constitute a link between enterprises and research institutes
with regard to technological demand and supply, but also to be able to
provide them with loans in order to shorten the lead time between R&D
findings and their commercialization.

4. To reform the existing public design institutes, and to set up engineering companies. Such companies would provide a connection between entrepreneurs, technology owners, machine builders and construction companies; they would pay attention to the selection of various
new technological discoveries; and they would then put them to use in
the design of manufactures or production lines. Their functions would
range from design contracting and the distribution of equipment to
providing turn-key plants, and would therefore play an extremely important part in the improvement of productivity.

5. To encourage R&D institutes and their scientific and technical
personnel to set up enterprises so as to gain immediate access to the
economy. This would be a beneficial change in the operational mechan-

isms of the development-oriented R&D institutes, and might facilitate
the establishment of a large number of small- and medium-sized enterprises in urban areas. It would also be useful for the self-develop-
ment of R&D institutes and, in consequence, for the improvement of the
living standard of scientific and technological personnel.

6. To channel engineers and technicians from university and R&D institutions into production, by means of creating part-time jobs,
short-term appointments, or research posts. This would provide an
appropriate adjustment of the distribution of S&T resources, as well
as relieving the situation in some enterprises that have been seriously short of technological personnel. This system would be much more
feasible than transferring people from one post to another.

7. Once production technology has been improved and the workforce is
better qualified, to make use of technological findings from abroad in
order to produce high-tech export commodities. The R&D workforces of
some developing countries and regions are not as strong as ours, yet
they are capable of exporting high-tech products. This is what we
should pay attention to.

8. To strengthen the engineers' and technicians' sense of the commodity economy. The priority must be shifted from academic values to
economic benefits, from the worship of discovery and invention to an
emphasis on various innovations that comprise little change in design
and production but are suited and beneficial to the market, and from
considering technical work as the only decent regular occupation to
becoming keen on selling, servicing, maintaining and other related
work essential to the development of a commodity economy.

9. To match advanced technology with excellent management. On the
basis of the current circumstances in China, it is much more difficult
to change the backward management than to overcome technological backwardness. Engineers and technical personnel should actively participate in business management and advance management reforms. It is impossible to attain a higher degree of efficiency and productivity if

attention is only paid to "hardware" but not to "software". Moreover, management improvement can sometimes result in higher profits at lower costs.

10. Technology should be developed in coordination with the economy and with society. Technology plays a great role in the promotion of economic and social development, yet this role must not be overestimated since technology can only be developed with economic and social support. It is desperately necessary to draft a complete policy on the intensification of the reform of the S&T structure, to create a social environment which will encourage technological progress, and to advocate studies on the correlation between technology and society.

Of course, the mechanisms of the commodity economy are by no means a panacea. Rather, we must try to explore various ways and methods to accelerate technological progress.

The 1990s will be the important decade for Chinese socialist development. There might be many more difficulties and obstacles than anticipated; we must be fully prepared in this respect. As long as we base our efforts on China's realities, however, as long as we adhere to reforms and open policies, as long as we firmly rely on the people, Chinese productive resources will be bound to achieve a rapid development, and the Chinese standard of living will experience a great improvement.

THE RAPID EXPANSION OF ECONOMIC INFORMATION IN THE 1990s
AND THE CHALLENGE TO CHINA'S ECONOMIC REFORM

Zhou Xiaochuan
Assistant Minister of Foreign Trade
and
Yang Jianhua

In the early 1980s, people began to talk about the emergence of a new technological revolution, and described it as the "third wave", maintaining that developed countries (DCs) had entered the "information society", "post-industrial society", etc. Some economists regard this change as a "megashift".

China began to discuss the new technological revolution at a fairly late date, and yet this belated understanding still exerted a notable influence on the country's economic theories. In a speech delivered in October 1983, the former Chinese Premier, Zhao Ziyang, mentioned the book, The Third Wave, by the American futurologist, Alvin Toffler, and another book, Megatrends - Ten New Directions Transforming Our Lives, by the American economist, John Naisbitt. Zhao stressed the importance of studying the new technological revolution emerging in the world, as well as China's countermeasures. Later, heated discussions about, and extensive studies of, that revolution began throughout China. It is, however, a pity that China, which is far behind other countries in terms of the adaptability of its economic system and economic-cum-technological structures, has failed to turn its attention to the "third wave" into specific policies, with the result that the upsurge has gradually cooled down. Nevertheless, it is still true to say that the discussions and studies at that time greatly broadened the vision of Chinese academics.

Today, people have begun to show an interest in, and to discuss, the world economy, which is expected to undergo drastic changes in the 1990s, and its impact on China's modernization. It can be said that on the basis of previous discussions and studies, people have acquired

Europe-Asia-Pacific Studies in Economy and Technology
Leuenberger (Ed.) From Technology Transfer
to Technology Management in China
© Springer-Verlag Berlin Heidelberg 1990

a new understanding in various degrees, and new studies have been made. This paper tries to expound, by proceeding from China's realities, major changes brought about by the new technological revolution, as well as their influence on the measures to reform China's economic system in the 1990s. In other words, theoretical analyses of China's economic reforms have so far still basically been limited to such traditional fields as decentralization, market regulation and ownership, while neglecting the new subjects concerning economic theories and systems analyses related to the rapid expansion of information.

1. THE RAPIDLY CHANGING WORLD ECONOMY

The rapid and yet imperceptible development of the new technological revolution will exert an immeasurable impact on the world economy in the 1990s. The industrial structure of the world economy will undergo changes at a faster pace. The proportion of the traditional primary and secondary industries will diminish further, while the tertiary industry focussing on the information sector will grow vigorously. As the demand structure moves more notably toward diversified and personalized patterns, the product mix of the manufacturing industry will be subject to more marked changes.

Various countries will participate more actively in international specialization, thus stepping up international economic and technological cooperation. With a further change in the patterns of economic strength in various countries, the economic status of countries and regions in the Asian Pacific basin will continue to grow, and LDCs (developing or less developed countries) will also undergo changes affecting their relative advantages. With a constant improvement in the industrial structure, and with the emergence of protectionism in DCs, LDCs will have a diminishing comparative advantage with regard to their primary and labor-intensive products.

All these tremendous changes will not only present a serious challenge to China's development prospects, but also a rare opportunity. A correct understanding of the changes is vital to China, which is carrying out reforms, pursuing open policies and generally modernizing

itself, because it will help China recognize the economic and technological gap between itself and the rest of the world. This challenge
will further help China to consider, on the premise of its economic
system reforms, its own economic development strategy for the 1990s,
particularly the assessment of the feasibility of the proposals coming
from different reformer schools - one of them stressing the urgent
need for price reform in the coming years, another advocating only
enterprise reform while dodging price reforms for the time being.

(1) A shift from the traditional primary and secondary industries to
the tertiary industry focussing on the information sector

Economists have expressed different views on the prospects of the
world economy in the 1990s. Some hold that a new economic crisis will
appear, while others maintain that, to the contrary, the 1990s will
enjoy a new long period of economic prosperity. Most economists, however, agree with the prognosis that the industrial structure of the
world economy will undergo a notable change, i.e. that the primary
sector (agriculture) pivoting on material production, and the secondary sector (the processing industry), will decrease in terms of their
proportion to the total GNP, while the tertiary sector focussing on
the information sector will continue to grow vigorously. The information sector has grown to such an extent in the United States that it
accounts for a considerable portion of that country's GNP. The rapid
development of this sector is related to the fast growth of inform-
tion processing, electronics and communication technology. Statistics
show that people employed in the information sector now make up more
than 75% of the national working population. With the rapid development of the information sector, the speed for building up new enterprises in the United States is at least seven times as high as in the
1950s, and the amount of information is now several times as much in
volume as it was a few years ago. Information, knowledge and technology have become the most important wealth and resource factors for
production in modern society.

(2) <u>A marked change in the structure of the value added in traditional</u>
   <u>industries</u>

In the 1990s, the consumption pattern will include a large part related to information and culture, while there will be a marked increase
in the value added created by the handling of a variety of information
in the course of design, production, management and marketing. Their
main expressions will be:

* The consumption pattern will move rapidly toward personalization
and diversification. New designs, styles and varieties will be
sought, while the ratio of "software" and its carriers will grow considerably.

* Advertising, trademarks and intellectual property will play an increasingly important role.

* Research and development will become increasingly important.

* Product design will increasingly depend on data retrieval from data
bases and on working with CAD systems, so that the ratio of the information-related value added of the design work to the total product
value will rise markedly. The relationships between the design of
consumer goods and the evolution of world culture, particularly the
relationship between new cultural products and new cultural trends,
will become increasingly closer.

* Large amounts of special-purpose equipment and a whole range of
automatic equipment will keep emerging. Producers of traditional
small series will also find themselves unable to resist this change.

* In order to meet the needs for production in smaller quantities or
in more flexible forms, CAM, robots, FMS and combined cutting machines
will be involved on a larger scale.

* In the field of those information media which turn design into
production, many new methods have been adopted so as to be more

accurate, effective, and capable of delivering on time. NC, CNC and CAM have linked design more closely to production.

* In order to make production management meet the requirements of quick change and intense competition, the management of inventory, transport, finance and personnel is undergoing a fast modernization. In consequence, investment in human resources and the training of managerial staff and technicians have become increasingly important, causing costs, however, that cannot be disregarded.

* In the field of marketing, there is a series of well-known informa- ation services whose ratio is rising. Most profits made from product sales do not now come from the manufacturing industry but from sales services and from trade. The growth of the public relations sector is in step with that of the information media.

To sum up, the ratio of the value added in relation to the above in- formation services continues to rise, with high technology and the service industries penetrating into the traditional industries, while the value added by unskilled labor is diminishing. This is why the United States and other developed industrial nations are paying enor- mous attention to the deregulation of the service sector.

(3) The sustained development of high-tech and knowledge-intensive industries

High-tech industries will grow faster in the 1990s. IC and LSIC pro- duction, for instance, is shooting up while the selling prices are plummeting, and this trend will continue.

Great changes have taken place in the composition of costs for elec- tronic data processing. The ratio of software costs to hardware costs was 1:9 in 1950, becoming 8:2 by 1985. The widespread use of compu- ters and electronic technology in such areas as industry, national defense, scientific research, education, administration, commerce and medical science will greatly enhance efficiency and productivity in these areas, and promote research on and the development of high technology, including biological engineering, marine technology and space technology. In the 1990s, the production structure will undergo more rapid changes; the manufacturing industry will have a relatively smaller demand for natural resources, and the ratio of labor costs to total production costs will become relatively smaller, whereas the ratio of costs for techniques and high technology (e.g. R&D costs) will rise rapidly. In the wake of the development of high technology, market competition will become even fiercer, making it imperative for enterprises to acquire easy access to information, respond promptly to market changes, and be quick in making decisions and strive for a high degree of efficiency. Therefore the organizational forms of enterprises will further grow in two directions: on the one hand, toward large-scale enterprises and transnational companies, and on the other hand, toward small- and medium-sized enterprises that are highly flexible and fairly efficient, constituting the overwhelming majority of enterprises. Large corporations achieve economies of scale through mass production and employ modern information technology to establish a planned management that used to be difficult to implement but now reduces unnecessary internal transaction costs and controls sales channels on the international markets, while SMEs meet diversified, specific and constantly changing market demands.

(4) Changes will take place in the comparative advantages of LDCs

Restricted by the level of their economic and technological development, LDCs mainly export primary products, and finished and semi-finished labor-intensive products. At present, LDCs may compete, on the strength of their advantages in abundant primary resources and low labor costs, with DCs on international markets. Their export-oriented development strategy constitutes a short-cut for their primitive capital accumulation and for the modernization of technology and management. Nevertheless, owing to the rapid growth of the new technological revolution and the concomitant drastic changes to the industrial structure, profits made by exporters of primary and labor-intensive products from LDCs are dropping, and this trend will continue through to the 1990s.

The large-scale research on and development of energy conservation and material-efficient technology has meant that plastics, synthetic leather, synthetic fibers, blended fabrics, glass fibers, high-grade ceramics and other new materials have become substitutes for primary resources including cotton, wool, jute, steel, copper and natural rubber, thus slowing down the growth in the demand for primary products on the international market and causing a sustained drop in their relative prices. Between 1980 and 1985, prices of primary products on the international market dropped by 32%. An estimate by the World Bank shows that in the 1990s, the market demand for non-fuel primary products will grow at an annual average rate of only 2.1%, while their relative prices will continue to drop at an annual average rate of 1%. The ratio of primary products to international trade has steadily dropped to only 40% and will continue to do so in the 1990s.

Asian NICs have achieved an enormous economic growth by making use of their export-oriented manufactured goods at low labor costs. This model will draw close attention from LDCs, and yet changes have taken place quietly and gradually in international market conditions, resulting in a big drop in the growth rate of the world trade volume as compared with 20 years ago. Profits made from the exports of labor-intensive products have become smaller, and the impact produced by the exports of such products on primitive capital accumulation has become weaker. The new trend of the world economy in the 1990s will require LDCs to reform their economic policies and improve their adaptability in order to build export-oriented industries and attract foreign investment. For this reason, the economic development strategy should be reassessed and discussed again.

(5) The interdependence of, and clashes in, the world economy will become more intensive

The 1990s will see an intensified trend toward the internalization of economic activities in various countries. International practice has indicated that economic growth, technological process and innovation can only be promoted through international exchanges, specialization and competition. Since the DCs have some similarities in their economic and technological structures, they will intensify and promote economic specialization and cooperation with one another. In DCs, transnational companies, in their capacities as global producers, traders, investors and technical inventors, will play a greater part in promoting cooperation in the form of further specialization and competition. The government of these countries will improve "macro-economic" co- ordination and intensify joint interventions in the field of international finance.

In the meantime some LDCs will also open themselves wider to the rest of the world economically, participate more actively in international specialization and competition, and link their economies more closely to the world economy. They will expand both mutual economic relations and economic and technological ties with the DCs. DCs, however, may possibly enforce stricter protectionism on labor-intensive products from LDCs. The growth rate of the world trade volume might be kept at a rather low level, making it more difficult for LDCs to expand their export sectors and making greater requirements on their import-export structures. Owing to the differences in their economic structures and development strategies, LDCs are therefore very likely to show a further dichotomy in their development performances, which will result in the trend toward further polarization among them.

2. SOME INDICATIONS TO BE OBSERVED IN CHINA

While discussing the new technological revolution around 1983, a great many Chinese academics and officials agreed that China's production and consumption levels were so low that the country would not be seriously affected by the third wave, so that China should continue to emphasize the development of the traditional sectors while simultaneously attaching due importance to new changes. A sober analysis will show that the changes have gone far beyond what those people anticipated five years ago, and if we paid no attention to these indications, then we would continue to underestimate the necessity of countermeasures in the rapdily changing 1990s.

Comparative studies of China's consumption pattern used to employ the concept of the per-capita income in order to forecast an evolution in consumption. Looking at it conscientiously, we will find that such studies sometimes contained grave errors. Consumption at a low level is still subject to the influence of cultural trends in the world, seeking personality and diversification, as for instance in the development and the success of fashionable articles of clothing. The sales of consumer durable goods related to new technology, particularly household electric/electronic appliances, are very good. A considerable part of the sales is hardware or software related to information exchange and culture. The sales of some durables are more than ten times the amounts previously predicted for the Sixth Five-Year Plan Period (1981-1985).

Another unexpected change is that China now has 200,000 computers of various kinds, including PCs. Although some are not used very efficiently, the figure far surpasses conventional forecasts.

In the export-oriented processing industry, China now achieves a very low value added and makes very low profits from cheap labor, which calls for serious attention. FOB prices of export processing products are normally one fourth to one fifth of the final retail prices in the DCs, leaving their value-added part to information-related fields including sales promotion, wholesale, retail sales, advertising, trademarks and style design - fields in which Chinese manufacturers have great difficulty in becoming involved. Also, Chinese processing industries must be capable of satisfying the final customers' differing tastes, of keeping pace with market changes, of ensuring product quality and safety standards, which is why they must usually use DC-supplied special equipment, molds, patents and technology, training facilities, and design and managerial expertise, all of which results in costs. As a consequence, primary labor usually contributes about half of the one fourth or one fifth of the value added, which corresponds to between only one eighth and one tenth of the total output value. In the coastal Fujian Province, for example, the payment for assembling one quartz watch is less than two US cents; of course such earnings are regarded as good by these Chinese TVEs (town and village enterprises), but we should not ignore the enormous share of the value

added created by the information related service industries. If China
continued to lag behind in the field of information, it would hardly
be able to reach the goal of fast-expanding export gains based on low
labor costs and abundant natural resources, using export commodities
as their vehicles, but would be highly likely to become a small con-
tributor - accounting for only one tenth of all value added - to
business operations largely based on information processing in DCs.

In China, many typical export-oriented, labor-intensive processing
industries are undergoing changes, and a great change is taking place
in the concept according to which TVEs stand for labor-intensive production. Zhao Ziyang, the General Secretary of the Chinese Communist
Party Central Committee, was surprised to notice that quite a number
of TVEs had so much, and such advanced, imported equipment. The point
is that the world markets, particularly the markets of DCs, set high
standards, which in turn are diversified and quick to respond to
change; also, big western companies have been controlling, at least to
some extent, consumers' taste, quality requirements and safety standards. In consequence, China's export-oriented manufacturing industry
has to buy equipment, tools, molds and even raw materials from DCs or
NICs for export production, while the domestic equipment-supplying industry usually acts so slowly as to lose good opportunities, even if
it were capable enough. The imported equipment, tools and molds, as
well as personnel training, are so expensive that the capital contribution rate in the production function of these industries is rising
substantially, while the contribution rate of primary labor is dropping sharply, so that people wonder whether they are still working in
labor-intensive industries and, in particular, whether they will undergo a further evolution in the 1990s. In principle, of course,
labor-intensive industries will not vanish, and there will still be
opportunities for a revival; but we cannot ignore the obviously
necessary change in the value-added structure any longer.

In traditional Chinese drawn-work, for instance, as well as in embroidery and other Chinese arts and crafts, the proportion of machine-
embroidered and machine-made products is already very high. The
machine-building plant in Qingdao uses highly complex special-purpose
equipment made in the Federal Republic of Germany. Computers are used

design products, to make prototypes, to check and control special-purpose equipment with program paper tapes. This involves very little primary labor, and most personnel are concerned with prototype design, computer and control medium operation, maintenance of special-purpose equipment and other information-related work. The embroidery sections of many export-oriented clothing and hat factories have introduced multihead embroidery machines on-line/off-line controlled by computers. If they had not done so they would not have met the required productivity or quality standards or been able to cope with the fast changes in style.

At present, Chinese manufactured goods are adopting a profusion of western company trademarks and prototype designs, while at the same time the marketing channels of those western companies are to some extent controlling consumers' tastes, product quality and safety standards, and therefore the production processes of the Chinese exporters. Bamboo chopsticks exported for use by overseas Chinese or Chinese restaurants in the US are a traditional Chinese product, but since sellers set high requirements on the chopsticks' straightness and burr, and therefore have engineers develop special equipment, only chopsticks processes with such equipment may be exported to the United States. It can be inferred that special-purpose equipment based on know-how and information is all-pervasive.

Consumption patterns displayed by DCs have also had a certain effect on the consumption pattern in China, which has a per-capita income of only 300 US dollars. The above-mentioned phenomenon in the processing industry does not only occur in exporting industries but also in manufacturing industries meeting domestic consumer demands.

In the joint ventures between Chinese and foreign companies and in enterprises wholly owned by foreign companies, the weak information industry on the Chinese side also brings losses. Chinese enterprises are weak where their own trademarks' industrial property, product quality and safety standards are concerned, with the result that transnational companies have opened up markets in China by installing assembly lines in the form of joint ventures or wholly foreign-owned enterprises, whereas the value added by their Chinese partners is extremely

low. Some foreign firms are inclined to move to China those industries which are high on energy consumption but create little value
added, leaving at home those industries with a low energy consumption,
a high information value and a large value added; this puts excessive
pressure on China's infrastructure, energy resources and environmental
protection. China is backward in terms of the systems concerning intellectual and industrial property. Superficially it has the advantage of spreading technology, but it still often suffers in the field
of trade and will be in a disadvantageous position in future international specialization and service deregulation.

3. THE CHALLENGE OF THE REFORM OF CHINA'S ECONOMIC SYSTEM

With the development of the new technological revolution, the world
economic pattern in the 1990s will undergo further changes which will
present China with both a rare opportunity and a serious challenge.
First of all, China is far behind the advanced world standards in economic as well as technological terms. Its industrial structure is
distorted and irrational, and some of its industrial sectors, especially the service sector, have been neglected for a long time.
Again, some other industrial sectors have grown excessively, thus
seriously limiting national economic growth as a whole. The proportion of China's agriculture and raw materials industry is fairly
large, about 33%, compared with around 10% in many other countries and
3% in DCs. The proportion of China's tertiary industry, which has
grown slowly, is only 20%, compared with an average of 61% in DCs and
67% in the United States. China's secondary sector accounts for the
fairly large proportion of 47%. In comparison with countries that
have the same per-capita GNP, China is quite advanced in terms of industrial equipment, technology and expertise. In terms of the scale
of fixed assets in the machine building and electric/electronic industries, it even holds a relatively high position among countries with
per-capita GNPs ranging from 300 to 4,000 US dollars. But because of
the slow development of its infrastructure, including energy and
transport, and because of its "soft" conditions such as administration, mobility, education and personnel training, China's manufacturing industry has a relative surplus in production capacity. Moreover,

a distorted price system, high exchange rates, a rigid control of resource allocation, and restrictions on the Chinese enterprises' autonomy in both domestic and foreign trade, all make for a considerable
loss of efficiency in the allocation of China's scarce resources,
making it impossible for production factors to flow in reasonable directions and thus enabling a full use of the surplus production capacity. This is why China is confronted with the arduous task of read-
justing its economic structure in the face of the world economic
patterns of the 1990s.

In order to answer the question of how to improve the industrial
structure and how to promote a higher degree of efficiency, Chinese
economists have begun to attach great importance to the study of market mechanisms and theories, which has led to heated discussion of
macro-control methods, enterprises' decision-making power, and ownership. On the one hand, practice has shown that this always goes contrary to planners' wishes, since they are trying to improve economic
structures and to activate the potential capabilities of enterprises
by means of mandatory plans. As a result, people have come to realize
the necessity of attaining this objective by further relying on market
regulation and a rational price system. On the other hand, the rapid
development of new industrial sectors calls for the establishment of
flexible and effective enterprise operations, which presupposes a reform of the distorted price system, the implementation of a competitive market system, and the achievement of an optimal allocation of
resources, as well as their optimal utilization. It is therefore absolutely necessary and urgent to call for a coordinated package of reforms for the price system, the fiscal and taxation system, the banking and trade system, with the aim of instituting a unified, competitive market mechanism. This, indeed, is the key question: will the
China of the 1990s be able to achieve its modernization goals through
technological progress and the mechanisms of competition, thus narrow-
ing the gap between itself and DCs? If China fails to make a breakthrough in economic reform and promptly to implement socialist market
policies in line with the trend of the world economic development,
then the rapid changes affecting the world economy will present China
with a challenge rather than an opportunity.

Nevertheless, some Chinese economists insist that China may dodge the price reform and the establishment of a market system, and instead start reforms in some easier, less risky fields. They share the view that although pricing and market mechanisms, and their function of regulating the relations between demand and supply, are important, it is not true that there is no alternative transitional approach to this problem.

One issue this paper intends to emphasize is that among Chinese economists there has always been a weak subject - an understanding of the great significance of the economic information structure, of information exchange, and of their cost in the economic system; an assessment of the interconnections between the information system and the market system, and between the price system and industrial policies; and a consideration of reform targets for different periods from the point of view of the economic information system. If the economic situation of the 1990s is really characterized by a rapid expansion of information, then the information characteristics of China's economic system will become a crucial issue - reform designers neglecting the significance of economic information will probably have worked in vain.

Another issue this paper intends to emphasize is that if China wants to compete in the world economy of the 1990s, which will be characterized by constant innovation and rapid change, then it must institute a new system to encourage innovative activities including those in the field of appropriate technological inventions based on effective information. This calls for a system that is capable of correctly assessing various innovative activities, including those in the service industry.

This paper maintains that attention should be paid to solving the following problems in the existing economic system:

1. The backward information structure with traditional CPE features and a corresponding information exchange system. In an era characterized by an information explosion and the use of information processing as a major approach to the creation of value added, effective informa-

tion exchange methods and the setting-up of a reasonable information structure have become a question of vital importance. At present, comparative studies of the performance of the economic system have shown that market policies incorporating an effective macro-economic control have proved successful, with the price system serving as a coordinating medium for the exchange of information in an economic system of decentralized decision-making. It enables micro-economic entities to forge horizontal ties and minimizes the costs for acquiring information without virtually any unnecessary delay. It has also been demonstrated that the CPE does not suit an economy based on mass-information because it involves high costs, rather large losses and a certain amount of distortion, as well as long delays in information transfers from grass-roots micro-economic entities to top authorities and _vice_ _versa_, with the government usually lacking the ability to operate such complicated information systems. It is therefore imperative to institute an appropriate information system in order to normalize the behavior patterns of enterprises and other organizations. The legal system which regulates corporate and organizational behavior patterns must be organically integrated with an appropriate information structure. Price, tax and legal systems should therefore also serve as information packages which would put a check on the behavior of enterprises, coordinate various interests, and guarantee an efficient and flexible resource allocation. Moreover, since all kinds of economic signals - including prices, the exchange rate, and interest rates - serve as standard parameters for the evaluation of economic efficiency, any error in these signals and their transmission will result in heavy losses, such as a decrease in the market share, mistakes in the assessment of project feasibility and of decisions with respect to economies of scale. The CPE system and the two-tier price system during the transitional reform period in China will distort price signals and their transmission channels, resulting in intransparency and a loss of market opportunities, as well as problems with income distribution and high costs arising from illegal "public relations".

2. Enterprises lack autonomy in management. To be in line with an ever-changing information structure, enterprises must be able to be flexible in their operations, make comprehensive arrangements and have

the benefit of a fully decentralized decision-making power. To date, many Chinese enterprises do not have the authority to fix the prices of their own products and to sell them, which may be regarded as the most important managerial decision-making right; instead, they are subject to the formula of government list prices. Only when enterprises have the benefit of full decision-making authority in all market operations will market and price mechanisms be able to play their role in the optimization of resource allocation. The fact that some enterprises are not entitled to fix their prices means that other enterprises cannot have free access to markets and select various inputs on the market at market prices, which in turn means that they have to depend on allocations and administrative offices at a higher level. Moreover, since the labor and capital markets are closely related to the commodity market, in fact depend on the normal existence of the commodity market, many enterprises do not have any decision-making power with regard to the management of production factors, either. This lack of basic incentives has seriously impeded various innovative activities. The most typical example is that the current enterprise mechanism cannot ensure the return of investment in R&D on new products.

3. Poor competitiveness: technologcial progress and innovation require adequate competition. The development of the service sector will also require the deregulation of service prices, as well as competition. At the moment, the service sector is still growing fairly slowly, since price controls in this sector have not yet been completely lifted, nor competition been fully introduced. A lack of competition and the market segmentation caused by administrative decentralization will still provide opportunities for monopolies. At present, quite a lot of Chinese enterprises are still not allowed to get involved in foreign trade and international competition, while many trade channels are controlled by Hong Kong or Macao firms, which take away a considerable portion of the profits.

4. Insufficient emphasis on intellectual and industrial property. Technological progress and innovation largely depend on promotion and safeguarding intellectual and industrial property. For this reason, intellectual and industrial property must have a market protected by

law, as well as reasonable prices. In order to be able to achieve this, it is imperative that enterprises have the right to fix prices; otherwise, there will be no yardstick against which the selling prices of property rights can be fixed, and no incentives and motivation for technical innovation, so intellectual and industrial property will be devalued.

5. In some deregulated fields, market parameters show great instability. In a new situation marked by diversification, complexity and an increase in business uncertainty, a good economic system should develop a mechanism capable of absorbing and alleviating such insecurities. A market-based insurance and futures market will be an absolutely necessary means to this end. Since only a small part of the industrial material market is deregulated, and since the two-tier price system is in effect in China, frequent readjustments in the State plans will exert an indirect influence on the prices on the free market, thus increasing instability. In these conditions, enterprises will tend to be short-sighted, considering that under such circumstances it will not be necessary to innovate, that inaction is preferable to action, and that there is no point in new learning processes. This is largely a consequence of problems inherent in the economic system.

All these aspects do not provide an overall picture but show that the price system and the pricing autonomy of enterprises are closely related to the exchange of economic information and to the enterprises' initiative spirit in competition.

We do not think that market mechanisms and market clearing prices can solve all problems; after all, everyone knows of so-called market failures. Yet the price reform and the establishment of a market system will remain the top-priority questions for China to settle in its current economic reform and its preparation for the economic challenge of the 1990s. It is very difficult to find or invent a new instrument to substitute the comprehensive functions of the market and of market prices, although some Chinese economists have always tried to find a substitution and to avoid price reform.

4. THE CHOICE OF ECONOMIC REFORM TO MEET THE CHALLENGE OF THE 1990s

It is both necessary and urgent to accelerate the reform of the economic system. This paper suggests that China's economic reform should
not dodge the reform of the price and market systems, otherwise it
would lose many opportunities in the 1990s. The essential question in
economic reform, and the major obstacle, is still the establishment
and further improvement of the market system, and the key problem in
this context is still the price system and the enterprises' right to
fix their own prices. Another reformer school in China has merely advocated pushing ahead with the internal reform of enterprises, while
delaying price reforms. For this reason, a great many questions have
not been explained with any clarity, and it is also difficult to understand why that school has put great emphasis on the micro-economic
foundation of the economic system but then assert that the price question and the enterprises' pricing autonomy are a matter of macro-
economics. In fact, however, economic theory repeatedly informs all
students that price and pricing issues are at the core of micro-econo-
mics. We have also put special emphasis on the reform of commodity
prices: the establishment of a market system should start from the
establishment of a commodity market. Without such a reform, it would
be very hard to set up labor, capital and technology markets without
any difficulties, because wages, capital gains and technological gains
are partially derived from the relative commodity prices resulting
from specific factors that are put into the production of individual
commodities. The establishment of these markets will require both
coordinated efforts and an approximately rational arrangement of the
reform sequence. Some necessary risks inherent in such a reform cannot be avoided, but without substantial reforms there will obviously
also be risks, and bigger ones at that.

It is not easy to invent a completely new economic policy. Until now,
there have only been two economic systems in the world which are both
theoretically complete and inherently stable: the macro-regulated market economy and the CPE.

As to the market economy, it is successful and economical from the
point of view of the exchange of information and the decentralization

of decision-making; from the point of view of stability and fair income distribution, it should institute a market system with a regulatory macro-economic function and adopt different planning systems in some fields where prices cannot clear the special markets. In some DCs, business practices based on quasi-public ownership - in companies mainly owned by pension funds, insurance institutions, mutual investment funds, and public companies - show that the market system can also be used where public ownership holds a dominant share in the ownership structure.

The CPE sacrifices a certain amount of micro-initiative, though theoretically it can ensure the optimization of resource allocation, enable an active government involvement and fair income distribution; it is feasible and convincing if its information characteristics are not taken into account. In the contemporary world economy, however, the bulk of information is so vast and its changes so rapid that this presents CPEs with technical difficulties that they find hard to cope with. It is in the primary stage, where the amount of information is small, that the CPE system could acquit itself rather well.

Many comparative studies of the performances of the two economic systems in the last four decades since World War II, show that the CPE is in need of a certain amount of reform. Over the past few years, the experience of socialist countries and some LDCs has demonstrated that it is very difficult to put the two different economic systems in parallel and try to have the best of both worlds. It is also extremely difficult to invent a completely new economic system.

China's government, however, has already adopted a new model as the target for its economic reforms: "The government regulates the market, while the market gives guidance to enterprises." This paper would like to suggest that once this market-oriented reform on the basis of public ownership has taken firmer roots, the following reform measures should be adopted to meet the requirements of the world economy of the 1990s:

1. A reform of the price system, and the establishment of competitive market mechanisms. On the commodity market - which is the most ser-

iously distorted - the irrational price system may be reformed by means of methods whereby official prices would first be adjusted toward computed equilibrium prices, with deregulation to follow as soon as possible. In the meantime, the two-tier system should be abolished, and a market-clearing equilibrium system connected with world market prices should be adopted. Along with this reform process, it is an essential factor that in the sectors where commodity prices have become more or less rational, markets should be opened and gradually expanded in order to promote a reasonable mobility of production factors and to encourage competition and technological progress.

2. The institution of a new macro-economic regulation system to ensure the stability of the macro-economy, the integrity of the domestic market, and equal competition. In line with the needs of the reform process and of a macro-economic balance, China should consider employing indirect policy instruments to as great an extent as possible. The following reform measures may be suggested:

a. The planning system. A new planning system should focus on studying and drawing up a macro-economic development strategy and an overall balance. In order to ensure an organic connection with market regulation, the budgets of all plan-ensured key projects should use market prices as much as this is possible.

b. The financial and banking systems. Based on a rational distribution of responsibilities between Central Government and local authorities at all levels, China should adopt a taxation system whereby fiscal revenue is classified in certain categories and correspondingly divided up between Central Government and local authorities. Meanwhile, it is imperative to impose strict controls on government subsidies and fiscal deficits, and to stop financing the government budget by printing notes of the Central Bank. It is necessary to intensify the role of the Central Bank in the control of the money supply and of the scale of credit in order to maintain a reasonable aggregate demand. At the same time, commercial banks should be put out to market competition and made to adopt enterprise operation principles. The Central Bank should set and maintain a balanced exchange rate in accordance with the balance of payments.

c. To reform the taxation system. On the basis of the price reform, China should transform its current complex taxation into a standardized system consisting of a value-added tax and a corporation income tax at appropriate rates.

d. To speed up economic legislation.

e. To define the status of the enterprise as a legal entity, and to ensure enterprise autonomy in market activities. China ought to reform its enterprise mechanisms in many aspects. The enterprises' mannagerial rights should be resolutely confirmed, and it is necessary to abolish both the subordinate position of enterprises in respect of their administrative departments in local/central government, and the enforced sector-specialization imposed on enterprises, thus allowing them free access to both the domestic and the world markets. With regard to publicly owned enterprises, efforts should be made to improve their specific managerial forms in relation to public ownership, and to introduce widespread competition. Meanwhile, attention should be paid by Central Government to maintaining as best they can the balance between aggregate demand and aggregate supply, as well as the relative stability of conditions for the reform of enterprises.

5. CONCLUSION

The explosion of economic information expected to occur in the 1990s constitutes a serious challenge to China's reforms. This paper has highlighted some questions about this information explosion in a rapidly changing world, along with some basic ideas concerning the structure and exchange of information on a micro-economic level - ideas which might well be of some significance to China's reformers and planners looking ahead toward the world's and China's development in the 1990s. We are convinced that China will choose an appropriate reform strategy and achieve the model of reform objectives as defined in a document of the latest National Congress of the Chinese Communist Party: "The State regulates the market, while the market gives guidance to enterprises."

Contributors

Dr Manfred Kulessa, Chairman, Development Policy Forum,
German Foundation for International Development, Berlin

Dr Richard Conroy, OECD, Development Centre, Paris

Dr Pierre Ventadour, Directeur General, CODASIE (Compagnie
pour le dévelopment avec l´Asie de stratégies industrielles
et economiques), Lecturer at the University of Le Havre,
Paris

Ryusuke Ikegami, Institute of Developing Economies, Tokyo

Dr Jaques A. Astier, Consultant, Paris

Dr Eduard B. Vermeer, CHINATEAM Consultants, Leiden/NL

Prof Dr Lewis M. Branscomb, John F. Kennedy School of Government,
Harvard University, Cambridge MA

Oshima Keichi, Taiso Yakusiji Technova Inc, Tokyo

He Kang, Minister of Agriculture, Beijing

Hu Ping, President, National Research Centre, Beijing

Lin Zixin, Chairman, Science and Technology Daily, Beijing

Zhou Xiaochuan, Assistant Minister of Foreign Trade, Beijing

Yang Jianhua, Ministry of Foreign Trade, Beijing

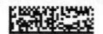